Sound control for homes

Building Research Establishment
Garston
Watford
WD2 7JR

Construction Industry Research and
Information Association
6 Storey's Gate
Westminster
London
SW1P 3AU

Prices for all available
BRE publications can be
obtained from:
BRE Bookshop
Building Research Establishment
Garston, Watford, WD2 7JR
Telephone: 0923 664444

Further details about CIRIA services and
publications, and subscription rates, can be
obtained from:
The Secretary, CIRIA
6 Storey's Gate
London SW1P 3AU
Telephone: 071 222 8891 Fax 071 222 1708

BR 238
CIRIA Report 127

BRE ISBN 0 85125 559 0
CIRIA ISBN 0 86017 362 3
CIRIA ISSN 0305 408X

Foreword

Sound control for homes presents practical, state-of-the-art advice on the control within dwellings of noise from outside sources and noise transmitted within and between dwellings. Its production was funded by the Department of the Environment through a contract directed by Dr L C Fothergill of BRE, and it was prepared by John Miller, acoustic consultant, under the supervision of Jeff Charles of Bickerdike Allen Partners.

Sound control for homes was first published by CIRIA in 1986 in two volumes, CIRIA Reports 114 and 115. This single-volume edition has been revised and expanded to include requirements for sound insulation in Scotland and Northern Ireland.

To ensure that the revised edition would meet the requirements of a wide audience, an advisory group was set up and chaired by Ann Alderson of CIRIA. The members of the group were:

Professor Peter Burberry, University of Manchester Institute of Science and Technology,

Mr A Chapman, Brighton Polytechnic,

Mr R H Clough, Wimpey Environmental Ltd, and

Mr A Holt, Holt Associates.

Information and comments were received from the following organisations and individuals during the preparation of the original volumes and of this revised edition:

Autoclaved Aerated Concrete Products Association,

British Gypsum Ltd,

British Cement Association,

Building Research Establishment,

Department of the Environment,
Building Regulations Division,

Department of the Environment for Northern Ireland,

Rendel Science and Environment,

Mr S Rintoul,

National House-Building Council,

Pilkington Glass Ltd,

Scottish Office,

Timber Research and Development Association.

Their contributions are gratefully acknowledged.

Contents

How to use this manual

If you are familiar with the principles of acoustics or have an immediate problem to solve, skip this section.

In parts B, C and D, certain words appear in CAPITALS. If you are unsure of their definition or interpretation, refer back to Part A for an explanation.

A glossary of terms is given on page 126.

This covers planning against external and internal noise sources and the selection of appropriate construction methods.

Turn straight to page 21 for directions to the topic of interest.

This gives details of construction relevant to sound insulation. All constructions which appear in the Building Regulations for England and Wales, Scotland and Northern Ireland, have been included. Turn to page 35 for further details.

Checklists covering design and site inspection are given with each construction type. For ease of reference, these are reproduced in appendices B and C, pages 121 to 124.

Sources of further information are also given for each construction type. The address of each organisation is given on page 128.

This provides worked examples which explain how the acoustic design methods given in parts A to C are applied in practice.

The examples fall into four categories:

Site noise assessment
Examples illustrate techniques used in the planning of a site for new dwellings and in the design of the building envelope to control external noise arising from a variety of sources. Appendix A gives details of the calculations used to assess noise on site.

Design of new dwellings
Examples illustrate noise control techniques used in the internal planning of various types of new dwelling and the selection of appropriate forms of construction.

Conversion properties
Examples illustrate design methods used to ensure reasonable sound insulation in converted properties.

Noise problems in existing dwellings
Examples illustrate methods used to solve noise problems in existing dwellings

Turn to page 77 for directions to the appropriate worked example.

Introduction

Scope

This manual is directed at architects but will be of use also to other building professionals. It covers the following aspects of acoustic design in housing:

- appraisal of noise affecting the site,
- planning to control external noise,
- planning to control internal noise,
- selection of appropriate forms of construction to control external and internal noise, and
- detailing for noise control.

The physical principles

Part A is a brief review of the physical principles of acoustics and noise control. Its inclusion is justified for a number of reasons:

- it introduces the uninitiated to the subject,
- it provides useful revision for those who were once familiar with the physical principles but need reminding of the main points, and
- it makes the manual self-contained, requiring minimal reference to text books.

Many users will have an immediate application for the manual and no need or desire to read Part A. They should turn straight to the beginning of Part B, using A for reference only. To assist in this, terms defined in Part A appear in capital letters in parts B, C and D. For ease of reference, a glossary of terms is also given on page 126.

The acoustic principles have been described primarily in words and pictures. Formulae, graphs, tables and calculation procedures are optional reading. They are generally found to the side of the main text. Their application is illustrated in Part D of this manual, 'Worked examples'.

Scheme design

Part B, 'Scheme design', will guide the architect during work-stages A to D from the Royal Institute of British Architects (RIBA) publication, *Architect's appointment*. Part B emphasises the practical planning measures which are available at project inception, and provides a basis for selection of construction methods for separating walls and floors, partition walls and floors, and the building envelope.

Detailing

Part C assists the architect in detailing his or her selected construction methods during work-stages E to G, and provides site inspection checklists for use during work-stage K, 'Operations on site'. All the constructions which appear in the supporting documents to the current Building Regulations for England and Wales, Northern Ireland and Scotland, appear here, though the scope of the manual is not limited to constructions which meet these Regulations.

Worked examples

The worked examples in Part D have been suggested by people of long experience in sound control for homes. They represent the most frequently encountered acoustic problems and deal with most practical situations, including the design and construction of new dwellings, sound insulation in conversions, and noise problems in existing dwellings.

To avoid constant cross-referencing, relevant charts and formulae from A, B and C have been reproduced in each worked example.

Further information

The appendices contain:

- details of the calculations carried out for the worked example which assesses site noise,
- designers' checklists and site inspection checklists,
- a method for calculating wall mass,
- a glossary of acoustic terms,
- names and addresses of sources of further information, and
- a short bibliography.

Accuracy

Wherever possible, sound-insulation values have been given for each form of construction illustrated. This information is often from a limited source of data and, even where the database is large, subject to variability. For example, nominally identical separating walls or floors, field-tested by the Building Research Establishment during national surveys, may give results which vary significantly. Mean figures are quoted, and it should be possible to achieve them if the advice on layout and detailing is observed. Inevitably, many of the quoted performance figures are based on a small sample. These are the best estimates available and should prove sufficiently accurate for design guidance. None of the quoted values is sufficient evidence in itself to demonstrate compliance with the Building Regulations (see page 16).

The calculation procedures have been devised for simplicity of use and cannot always be expected to give results accurate to the nearest decibel. An estimate of accuracy has been given where possible.

Quality of detailing and workmanship

The manual stresses the value of quality in construction to sound insulation. Quality embraces both detailing and workmanship. It must be emphasised that a failure in either can result in a construction that does not meet its expected sound insulation. As far as separating walls and floors are concerned, good detailing and workmanship are essential if the construction is to meet the Regulations.

Checklists for detailing and site inspection appear with every construction type in Part C.

Other design factors

It should be stressed that sound insulation is only one design requirement. There may be cases where sound-insulation requirements conflict with other requirements such as thermal insulation, fire safety, etc. Detailed advice on these other matters is beyond the scope of this manual. Designers must satisfy themselves that any construction fulfils all the other relevant design requirements.

Part A
Principles of acoustics and noise control

Readers who are already familiar with the principles of acoustics may prefer to skip this section.

In parts B, C and D, certain words appear in CAPITALS. Readers who require clarification of the meaning or implications of these words should refer to Part A, where they are explained.

A glossary of terms is given on page 126.

How are sounds described?

Sound is a form of energy which is transmitted through the air. In transmitting sound, the air particles vibrate and cause rapid cyclic pressure changes. Sounds are characterised by FREQUENCY and LEVEL.

Frequency

FREQUENCY is the rate at which the air particles vibrate. The more rapid the vibrations, the higher the frequency and the higher the musical pitch. 'Hum', 'drone' and 'throb' are words applied to sounds which contain mainly low frequencies. 'Whistle', 'squeal' and 'hiss' describe sounds containing mainly high frequencies.

Frequency is measured in hertz (Hz). Older books use the units 'cycles per second' (cps). The two are equivalent.

The human ear can detect sounds in the range 20 Hz to 20 000 Hz (20 kHz), approximately. Figure 1 illustrates this range of frequencies on a scale. The scale is not linear because the ear responds to proportional changes. For example, a doubling of frequency represents the same musical interval wherever it occurs on the scale. Thus, the 'equal' steps of the piano keyboard do not correspond to equal increments on the frequency scale.

For practical building purposes, a restricted frequency range is used. It is divided into octave bands and one-third octave bands. The term 'octave' comes from musical notation. It is the interval between the first and the eighth note in a scale and represents a doubling in frequency. Figure 1 highlights the range of octave and one-third octave bands normally used. Each frequency value has an associated wavelength. The higher the frequency, the shorter the wavelength.

Level

In transmitting sound, the air particles vibrate back and forth about a mean position. The further they move, the greater the energy of the sound. It is difficult to measure energy directly. It is more convenient to measure the magnitude of the fluctuating air pressure caused by the sound wave.

The SOUND-PRESSURE LEVEL (spl) is used to describe the magnitude of sound. The units used are decibels (dB or, strictly, dB re 2×10^{-5} pascals, which is the sound pressure selected to represent 0 dB). The formula is given in the box on the next page.

Figure 2 shows the relationship between sound pressure and sound-pressure level and illustrates how the use of a logarithmic scale compresses the numerical range to a more manageable one.

Typical noise levels of common sources are also shown in Figure 2; these levels should be taken as dB(A) values (see page 7).

The human ear responds to continuous sound sources in broadly the following way:

- A 1 dB increase is the smallest audible change in level. It would be noticed only if the two sounds were presented in quick succession.

- A 3 dB increase is the smallest audible change which could be detected over a period of time.

- A 10 dB increase represents a doubling in loudness to the ear.

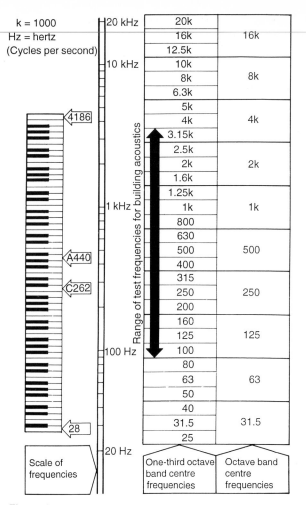

Figure 1

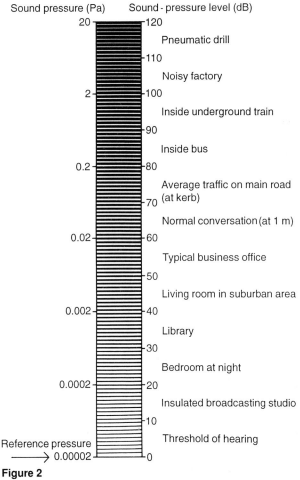

Figure 2

A sound can be described by reporting its sound-pressure level in each of a number of frequency bands. Information of this kind is normally obtained by measurement using a sound-level meter fitted with an octave or a one-third octave band filter set.

Sounds can also be described using a single-figure descriptor such as dB(A).

dB(A)

The ear notices high frequencies more easily than low frequencies. This characteristic has been incorporated into sound-level meters using the A-weighting network. Figure 3 is a graph showing the adjustments made at each frequency. This shape correlates well with human response to many types of noise.

A-weighted sound-pressure levels can be measured directly on a sound-level meter or obtained from octave-band sound levels by applying the following weightings at each frequency and then combining the resulting levels using the rules for addition of decibels (see box):

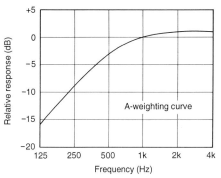

Figure 3

Sound levels that vary with time

One further factor must be considered in describing sounds, and that is the way in which they vary with time.

For example, free flowing traffic on a major road may be continuously audible, with fluctuations in level and frequency content related to individual passing vehicles. In contrast, aircraft may be inaudible for long periods between loud events. Research into the way humans are affected by noise has led to a variety of different measurement units for different noise sources (see pages 10 and 11). Two widely-used noise descriptors are defined here.

The equivalent continuous A-weighted sound-pressure level ($L_{\mathrm{Aeq},T}$)

Definition: $L_{\mathrm{Aeq},T}$ is the level of a notional sound which (over a defined period) would deliver the same A-weighted sound energy as the fluctuating sound.

The reference time period should always be stated. For example, eight hours' exposure to an equivalent continuous sound-pressure level of 90 dB(A) would be written: 90 dB $L_{\mathrm{Aeq},8h}$.

$L_{\mathrm{Aeq},T}$ values can be measured directly on many modern (integrating) sound-level meters.

Sound exposure level (L_{AE})

Where the noise over a given period is made up of individual noise events, the $L_{\mathrm{Aeq},T}$ can be predicted by measuring the noise of the individual events using the SOUND EXPOSURE LEVEL, L_{AE} (or SEL or L_{AX}).

Definition: L_{AE} is the level which, if maintained constant for a period of one second, would deliver the same A-weighted sound energy as a given noise event.

This unit can also be measured on some sound-level meters. The calculation procedure to obtain the $L_{\mathrm{Aeq},T}$ from L_{AE} measurements is given on page 11 under 'Railways'. Worked examples can be found on pages 114 and 118.

Sound-pressure level formula

$$\mathrm{spl} \;=\; 10 \log \left(\frac{\text{pressure measured}}{\text{reference pressure}} \right)^2 \mathrm{dB}$$

Notes

10	because the units are decibels. If the 10 were omitted, the units would be bels
( )²	because energy is related to pressure squared
log	because the ear responds more or less equally to equal ratio changes in energy. Logarithms express this mathematically. (Base 10 logs are used.)

Addition of dB

Sound-pressure levels are not combined by simply adding together the decibel values. A conversational voice at 1 m distance results in a sound-pressure level of about 65 dB. Two voices together would not cause a level of 130 dB; that would be louder than a pneumatic drill. Table 1 gives a simple rule for combining levels, and it produces a result accurate to ± 1 dB.

Table 1

Difference between levels to be combined (dB)	Add to higher level (dB)
0 or 1	3
2 or 3	2
4 to 9	1
10 or more	0

Example

If one source alone gives 65 dB, two identical sources together give:

    65 and 65 (difference 0, add 3) = 68 dB

Adding a third source gives:
    68 and 65 (difference 3, add 2) = 70 dB

Adding a fourth source gives:
    70 and 65 (difference 5, add 1) = 71 dB
OR 68 and 68 (difference 0, add 3) = 71 dB

To achieve the most accurate result when using this approximate method of combining more than two levels, start with the lowest number and work upwards.

Example for road traffic noise

A-weight and combine the following octave band sound-pressure levels measured at the kerbside of a main road to obtain a total level in dB(A).

Frequency (Hz)	Sound-pressure level (dB)		
		A	Combine
125	79	63	63 & 64 = 67
250	73	64	66 & 67 = 70
500	72	69	69 & 70 = 73
1000	72	72	71 & 72 = 75
2000	70	71	73 & 75 = 77
4000	65	66	Answer 77 dB(A)

How does sound travel in the open air?

Sound levels decrease with distance from the source. The rate of reduction with distance varies according to source type.

Two simple cases are considered.

Point source

The sound source can be represented by a point in space. Sound is radiated equally in all directions. Energy is constant for all points described by a sphere centred at the source.

The relationship between distance and energy is illustrated in Figure 4. The larger the radius of the sphere, the larger will be its surface area and the more thinly spread the sound energy passing through it.

- If the radius of the sphere is doubled, the surface area increases by a factor of 4 and the energy decreases by a factor of 4.

- If the radius of the sphere is increased by a factor of 3, the surface area increases by a factor of 9 and the energy decreases by a factor of 9.

- And so on.

This is known as the INVERSE SQUARE LAW.

In dB terms, a change in energy by a factor of 4 is 6 dB (= 10 log 4).

Every time the distance from a point source is doubled, the level decreases by 6 dB. Figure 5 shows this graphically.

(An aircraft in flight behaves like a POINT SOURCE as long as the observer is not too close.)

Line source

The sound source can be represented by a large number of point sources arranged in close spacing along an infinitely long straight line. The sound energy at an observer's position now relates to the area of a cylinder centred on the line. The relationship between distance and energy is illustrated in Figure 6.

- If the radius of the cylinder is doubled, the surface area increases by a factor of 2 and the energy decreases by a factor of 2.

- If the radius of the cylinder is increased by a factor of 3, the surface area increases by a factor of 3 and the energy decreases by a factor of 3.

- And so on.

In dB terms, a change in energy by a factor of 2 is 3 dB (= 10 log 2).

Every time the distance from a line source is doubled, the level decreases by 3 dB.

(Free flowing traffic on a busy road behaves like a LINE SOURCE as long as the observer is not too close.)

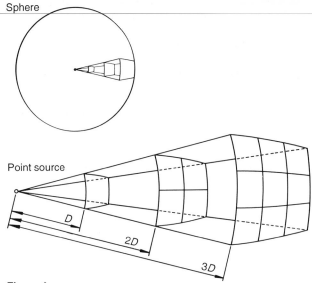

Figure 4

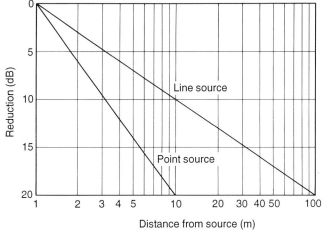

Figure 5

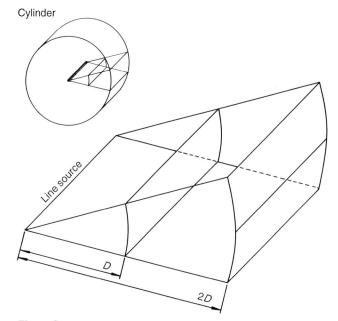

Figure 6

Surfaces and objects in the open air affect the way sound travels. The main factors which affect sound propagation follow.

Reflection and absorption
Principles

When sound hits an object, some of its energy will be REFLECTED and some ABSORBED.

Concrete, brick, timber and glass are good sound reflectors ('hard' materials). Their absorption coefficient is close to 0 over a wide frequency range.

Fibrous quilts are good sound absorbers ('soft' materials). Their absorption coefficient is close to 1 over a wide frequency range.

If a point source is located close to a hard surface, reflection causes all the sound energy emitted to be radiated away from the surface. Energy is constant for all points described by a hemisphere centred at the source (see Figure 7). All the energy passes through half the area of a sphere. The energy increases by a factor of 2 (or 3 dB) because of the presence of the reflecting plane. If the hard plane is exchanged for a soft one, the energy which would have been reflected will be absorbed, reducing the energy by up to 3 dB compared with the case with the reflecting plane. This effect applies to all types of source.

Sound propagation close to the ground

The presence of an absorbent plane close to the source does not generally affect the rate of reduction with distance. An important exception to this rule occurs, however, when the line between source and receiver is at a shallow angle or parallel to the absorbent plane. In practice, this would occur when the source and observer are both on the ground and there is grassland between them. The section on road traffic noise, page 10, gives information on the resulting attenuation.

Reflection from the facades of buildings

When noise levels are measured near to a building facade, the sound reflection from the facade will cause the level to be up to 3 dB(A) higher than would be the case if the facade were not present. Measurements made close to buildings (normally at a distance of 1 m to 2 m) are described as facade measurements and those made well away from reflecting surfaces are described as free-field measurements.

Barriers and diffraction
Principles

If a barrier is introduced between a source and an observer, blocking the line of sight, some sound will still travel to the observer. Some of the sound hitting the edge of the barrier is DIFFRACTED into the 'shadow zone' behind the barrier (see Figure 8). Nevertheless, barriers are effective noise control tools. For example, a well designed noise barrier can reduce road-traffic noise by approximately 10 dB(A) (see page 10).

Trees and hedges

Trees and hedges are not effective noise barriers. For example, a tree belt 100 m deep may reduce A-weighted road-traffic noise levels received on the other side by only about 3 dB(A) over and above the attenuation normally associated with soft ground. Consequently, tree planting is not a practical noise-control measure on sites for housing.

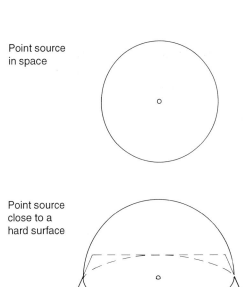

Point source in space

Point source close to a hard surface

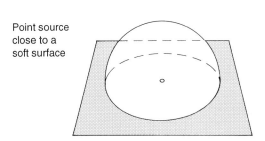

Point source close to a soft surface

Figure 7

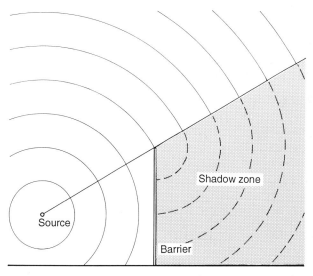

Source

Shadow zone

Barrier

Figure 8

External noise sources

Planning and noise

General guidance on planning and noise is given by the Department of the Environment (DOE)(see page 20). Where a specific site is under consideration, the planning department of the appropriate local authority should be consulted.

Road traffic

Criteria

The main regulations covering road-traffic noise are the Noise Insulation Regulations, 1975 and the Noise Insulation Amendment Regulations 1988, issued under the Land Compensation Act, 1973. These regulations do not apply to new housing (see page 20).

The object of these regulations is to compensate some residents subjected to additional noise arising from the use of new roads or roads which have been significantly altered. Road construction noise is also covered. If the additional noise is at or above a specified level, the affected residents receive a grant for double windows, supplementary ventilation and, where appropriate, venetian blinds and double or insulated doors. The specified level is 68 dB $L_{A10,18h}$, if there has been an increase of at least 1 dB.

> **Units**
>
> L_{A10} is the A-weighted sound-pressure level which is exceeded for 10% of a time period; for example, the level which is exceeded for a total of six minutes in an hour.
>
> $L_{A10,18h}$ is the arithmetic mean of the hourly L_{10} values for the 18 hours between 6 am and midnight on a normal working day.
>
> For most situations:
> $L_{Aeq,16h} = L_{A10,18h} - 2$ ($\pm$ 2 dB)

Factors affecting propagation

Reliable predictions of road-traffic noise can be made using *Calculation of road traffic noise* published by the Department of Transport and the Welsh Office. A site plan showing ground levels is required, together with the following data, which may be obtainable from the local authority's highways department:

- hourly traffic flow rate,
- mean traffic speed, and
- percentage of heavy vehicles.

It is usual to adopt predicted figures for 15 years hence. Charts are used to assess the noise level at any point on the site. Examples of the charts are shown in Figures 9 to 12. It is assumed that the source is situated 3.5 m from the kerb and 0.5 m above the road. The other factors which determine the final result are:

- road gradient,
- texture of the road surface,
- nature of the ground between the road and the receiver point,
- barriers and cuttings,
- angle of view of the road,
- reflections at the facade of the affected building (add 2.5 dB to the calculated value), and
- reflections from buildings opposite.

This method can be used to draw noise contours on a site plan. Worked examples can be found on pages 97 and 112.

If traffic conditions are complex or unusual, it may be necessary to measure the prevailing noise levels on site.

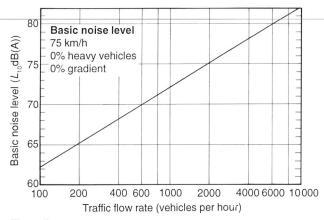

Figure 9

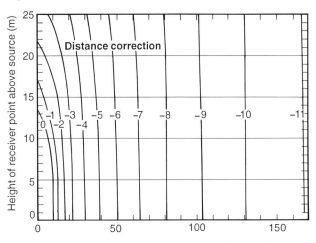

Figure 10

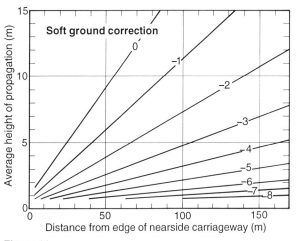

Figure 11

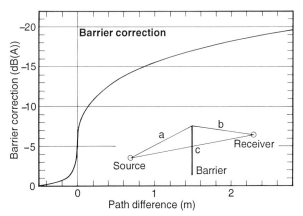

Figure 12

Aircraft
Criteria
A number of schemes pay grants towards the cost of sound insulation in existing dwellings in defined areas around major airports. The sound-insulation package is similar to that offered under the Noise Insulation Regulations for road traffic noise. The qualifying level has usually been 50 NNI, which is equivalent to 66 dB $L_{Aeq,16h}$ (see 'Units' in box). A similar scheme is offered around some Ministry of Defence (MOD) aerodromes for which the local authority hold $L_{Aeq,12h}$ contours. For new housing, individual planning authorities set their own policies with due regard to guidance issued by the DOE (see page 20).

Factors affecting propagation
None of the factors which determine the noise level on site is under the designer's control. For prevailing site conditions, reference should be made to noise contour maps available from the airport authority. If sound-insulation treatment appears necessary, noise measurements should be made on site to assess the maximum levels to which the site is exposed.

Railways
Criteria
Planning authorities set their own policies, with due regard to guidance issued by the DOE (see page 20). For example, the criteria in Table 3 were proposed by the Midlands Joint Advisory Council for Environmental Protection, which considered criteria adopted by a number of authorities. The Government has recently proposed that those responsible for new railways should be required to offer sound insulation where existing dwellings are exposed to a facade level of 68 dB $L_{Aeq,18h}$ (06.00 h to midnight) or 63 dB $L_{Aeq,6h}$ (midnight to 06.00 h). Noise insulation regulations for new railway lines are to be drafted.

Factors affecting propagation
There are no simple prediction methods for train noise. Levels fall off at approximately 5 dB per doubling of distance from the track over grassland. Road-traffic noise control measures are generally applicable also to train noise. As a general guide, train noise may be a problem if a site is within 30 m of a busy line without screening, and vibration may be a problem if there are lines up to 60 m away. A noise survey should be carried out if train noise is a possible problem. If groundborne vibration appears to be a problem, an expert should be consulted. (For criteria, see British Standard BS 6472.)

Units
Procedure for calculating $L_{Aeq,T}$

Measure the sound exposure level (L_{AE}), for each noise event such as a train pass-by. Note the values L_{AE1}, L_{AE2}, L_{AE3}, etc in dB. The $L_{Aeq,T}$ resulting from a given combination of noise events is obtained by inserting the measured values in the formula:

$$L_{Aeq,T} = 10 \log \left(\frac{\text{antilog } L_{AE1}/10 + \text{antilog } L_{AE2}/10 + ...}{\text{Total time period in seconds}} \right) \text{dB}$$

Worked examples can be found on pages 114 and 118.

Table 3

Day $L_{Aeq,15h}$ (07.00 h to 22.00 h)	Night $L_{Aeq,9h}$ (22.00 h to 07.00 h)	Action
< 55	< 50	No action necessary
55–65	50–60	Take action to control noise
> 65	> 60	No development recommended

Industry
Criteria
Planning authorities set their own policies with due regard to guidance issued by the DOE (see page 20). British Standard BS 4142 can be used to predict the likelihood of complaints. It offers a procedure to determine a rating level for the noise source in question (see 'Units' in box) and the value is compared with the background noise.

Factors affecting propagation
Industrial noise sources take so many forms that specific guidance on prediction methods cannot be given here. If an existing industrial noise is likely to be a problem, a site noise survey should be carried out, following the guidance given in BS 4142.

Units
$L_{Ar,T}$ Rating level

The noise level from only the noise source under investigation (the 'specific noise source'), is measured in terms of $L_{Aeq,T}$ and adjustments are made for the tonal and impulsive character of the noise. The resulting value, the rating level, is compared with the background noise, which is defined as the level exceeded for 90% of the time during the period for which the assessment is being made ($L_{A90,T}$). The reference time period, T, is 1 hour during the day and 5 minutes at night.

Rating procedure

If the rating level is 10 dB or more higher than the background noise, complaints are likely.

If the rating level is 10 dB or more below the background noise, complaints are unlikely.

How is sound insulation described?

One problem for the designer is that sound insulation is described in a variety of ways. This section should help in interpreting the data received from manufacturers. The most common terms are described in words. Formulae are set out in the box on the next page, and their use is illustrated by worked examples in Part D.

Methods for the measurement of sound insulation are laid down in British Standard BS 2750. This will be replaced in due course by a European Standard, which will contain extra parts and amendments. The latest edition should always be consulted.

When designing a home, sound insulation must be described in three main situations. In each, the observer is within the building.

- room-to-room airborne sound insulation (source is within the building),
- impact sound insulation (source is footsteps or other impacts on the building structure), and
- outside-to-inside sound insulation (source is in the open air).

Room-to-room airborne sound insulation

Voices, television sets, hi-fi and musical instruments are all sources of airborne sound (Figure 13). The airborne sound insulation between two rooms can be measured by generating a steady, broad-band sound level in one room (the source room) and measuring the sound-pressure level in both source and receiving rooms over a range of frequencies.

The LEVEL DIFFERENCE, D, is simply the difference between the source and the received levels ($L_1 - L_2$).

The level difference is influenced by the REVERBERATION TIME, T, in the receiving room. T is defined as the time taken for the level of reverberant sound in the room to decay by 60 dB. The STANDARDISED LEVEL DIFFERENCE, D_{nT}, makes allowance for the fact that most occupied domestic rooms have a reverberation time of about 0.5 seconds. Field measurements are standardised to this value.

The SOUND REDUCTION INDEX, R, is a property of the building construction only, independent of its area and the reverberation time of the receiving room. It must be obtained by laboratory measurements to ensure that the sample under test is the only element contributing significantly to sound transmission. (If this quantity is obtained from field tests, the result is called the APPARENT SOUND REDUCTION INDEX, R'.)

Impact sound insulation

The impact insulation (Figure 14) of a floor and its surrounding construction can be tested. The construction is excited by a standardised impact sound source (a tapping machine) and the resulting sound level, L_i is measured in the receiving room.

The STANDARDISED IMPACT SOUND-PRESSURE LEVEL, L'_{nT}, makes allowance for the fact that most domestic rooms have a reverberation time of about 0.5 seconds. Field measurements are standardised to this value.

The REDUCTION OF IMPACT SOUND-PRESSURE LEVEL, ΔL, is used in testing floor coverings. It is a measure of the improvement gained by laying the covering on a standard floor of 120 mm concrete. Figure 15 shows representative values for a variety of common floor finishes. The standard method of measurement may overestimate the value of thick carpets at low frequencies.

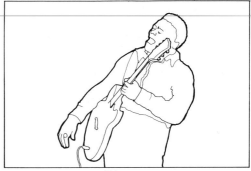
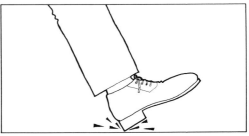

Figure 13

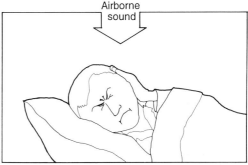
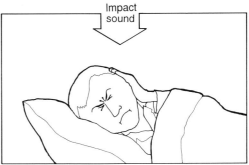

Figure 14

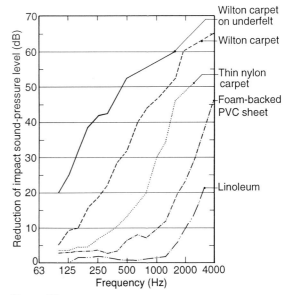

Figure 15

Outside-to-inside sound insulation
(Figure 16)

The sound insulation of a building facade can be tested using either road traffic or a loudspeaker as the source of noise.

Using traffic noise, the STANDARDISED LEVEL DIFFERENCE, $D_{nT,tr}$, can be obtained. Sound-pressure level measurements are made simultaneously outside the building facade and in the receiving room. The result is standardised to a receiving room reverberation time of 0.5 seconds.

The SOUND REDUCTION INDEX of a facade can also be obtained using road-traffic noise as the source. The value, R_{tr}, is independent of facade area and receiving room acoustics.

The SOUND REDUCTION INDEX of a facade can also be measured using a loudspeaker. The value, R_θ, is independent of the facade area and receiving room acoustics but is valid for only one angle of incidence (the angle between the axis of the loudspeaker and a line perpendicular to the facade). An angle of incidence of 45° is preferred.

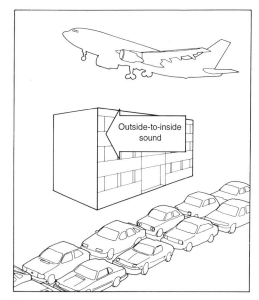

Figure 16

Formulae and symbols

A	Total absorption, receiving room (m²)
A_o	Reference absorption (10 m², a constant)
D	Level difference (dB)
D_{nT}	Standardised level difference (dB)
$D_{nT,tr}$	Standardised level difference, road traffic (dB)
L_1	Average sound-pressure level, source room (dB)
L_2	Average sound-pressure level, receiving room (dB)
$L_{eq,1}$	Equivalent continuous sound-pressure level 2 m in front of the facade (dB)
$L_{eq,2}$	Equivalent continuous sound-pressure level averaged in the receiving room (dB)
L''_1	Average sound-pressure level immediately in front of the facade but without the reflecting effect of the facade (that is, obtained away from reflecting objects) (dB)
L''_2	Average sound-pressure level in the receiving room (facade, loudspeaker test) (dB)
L_i	Impact sound-pressure level (dB)
L'_{nT}	Standardised impact sound-pressure level (dB)
L_n	Normalised impact sound-pressure level with a floor covering (dB)
L_{no}	Normalised impact sound-pressure level in the receiving room without any floor covering (dB)
ΔL	Reduction of impact sound-pressure level (dB)
R	Sound reduction index (dB)
R'	Apparent sound reduction index (dB)
R_{tr}	Sound reduction index (road traffic) (dB)
R_θ	Sound reduction index (loudspeaker) (dB)
S	Surface area of intervening construction (m²)
T	Reverberation time (seconds)
T_o	Reference reverberation time (0.5 seconds, a constant)
V	Receiving room volume (m³)
θ	Angle of incidence (degrees, 45° preferred)

Reverberation time

T can be measured in an existing room or calculated using:

$$ T = \frac{0.16 \times V}{A} \quad \text{seconds} $$

Conversely, the total area of absorption can be obtained if the reverberation time and volume of the receiving room have been measured, using:

$$ A = \frac{0.16 \times V}{T} \quad \text{m}^2 $$

Room-to-room airborne sound insulation

LEVEL DIFFERENCE, D

$$ D = L_1 - L_2 \text{ dB} $$

STANDARDISED LEVEL DIFFERENCE, D_{nT}

$$ D_{nT} = D + 10 \log T/T_o \text{ dB} $$

SOUND REDUCTION INDEX, R

$$ R = D + 10 \log S/A \text{ dB} $$

Relationship between sound reduction index, R, and standardised level difference, D_{nT}

$$ R = D_{nT} + 10 \log S/0.32V \text{ dB} $$

Impact sound insulation

STANDARDISED IMPACT SOUND-PRESSURE LEVEL, L'_{nT}

$$ L'_{nT} = L_i - 10 \log T/T_o \text{ dB} $$

REDUCTION OF IMPACT SOUND-PRESSURE LEVEL, ΔL

$$ \Delta L = L_{no} - L_n \text{ dB} $$

where $\quad L_{no} = L_i + 10 \log A/A_o$ dB (no covering)

and $\quad L_n = L_i + 10 \log A/A_o$ dB (with covering)

Outside-to-inside sound insulation

STANDARDISED LEVEL DIFFERENCE (ROAD TRAFFIC), $D_{nT,tr}$

$$ D_{nT,tr} = L_{eq,1} - L_{eq,2} + 10 \log T/T_o \text{ dB} $$

SOUND REDUCTION INDEX (ROAD TRAFFIC), R_{tr}

$$ R_{tr} = L_{eq,1} - L_{eq,2} + 10 \log S/A \text{ dB} $$

SOUND REDUCTION INDEX (LOUDSPEAKER), R_θ

$$ R_\theta = L''_1 - L''_2 + 10 \log \frac{4 \times S \times \cos\theta}{A} \text{ dB} $$

Refer to British Standard BS 2750 or the most recent edition of the replacement European Standard for further details of measurement methods.

What physical characteristics contribute to sound insulation?

Mass

The sound insulation of any single-leaf wall or floor built without gaps depends mainly on its MASS. According to the MASS LAW there will be an increase in sound insulation of about 5 dB if the mass per unit area is doubled (see Figure 17). Also, each time the frequency is doubled, a 6 dB increase should occur. This law does not hold good at all frequencies. A dip in performance occurs at the CRITICAL FREQUENCY (see Figure 18). This occurs when the wavelength of the sound in air coincides with the wavelength of the associated vibration in the wall. It is also known as the coincidence dip. This dip should not be allowed to occur in the frequency range of interest and certainly not in the range 100 Hz to 1000 Hz.

The critical frequency of a one-brick wall occurs at about 100 Hz, and of a half-brick wall at about 200 Hz. The sound insulation of the latter is poorer, not only because the MASS is reduced, but also because the critical frequency is at a point where its effects are more serious.

Isolation

In principle, double-leaf walls can give better sound insulation than single-leaf walls because the sound waves encounter two separate masses whose sound insulation could be added together. In practice, this could occur only if the two masses were completely ISOLATED from one another. Effective isolation has proved difficult to achieve in domestic masonry walls. For example, a 280 mm plastered cavity brick wall is no better than a 240 mm plastered solid brick wall, because of the higher critical frequency in each leaf and the bridging effect of the ties (see pages 36 and 40 for other comparisons).

Lightweight forms of domestic construction owe much of their sound-insulation performance to isolation. This can be illustrated by considering two sheets of plasterboard, each giving 25 dB sound insulation at 500 Hz. If they are bonded together, the mass law prevails and a result of 30 dB is obtained. If it were possible to ISOLATE them perfectly from one another, 50 dB would result (see Figure 19).

In practice, the ISOLATION will not be perfect. It depends on:

● the method of fixing (mechanical coupling) — how rigidly the two sheets are attached together, and

● acoustic coupling, which is affected by the width of the gap and whether there is an acoustic absorber in the cavity.

Figure 19 illustrates two practical compromises.

Effective isolation can also be achieved by forming a double-leaf partition from dissimilar leaves, for example masonry and lightweight panelling, because dips in performance in the two layers occur at different frequencies.

Extending the principles of the mass law suggests that the sound insulation of a double-leaf construction should increase by 10 dB with each doubling of mass, and by 12 dB with each doubling of frequency. However, the increases are much less in practice, because of acoustic and mechanical coupling.

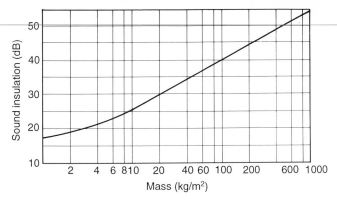

Figure 17

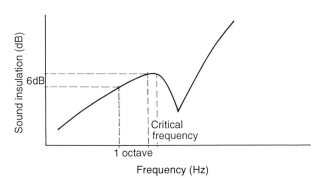

Figure 18

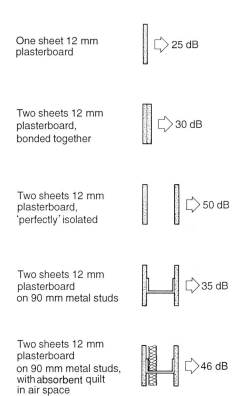

One sheet 12 mm plasterboard — 25 dB

Two sheets 12 mm plasterboard, bonded together — 30 dB

Two sheets 12 mm plasterboard, 'perfectly' isolated — 50 dB

Two sheets 12 mm plasterboard on 90 mm metal studs — 35 dB

Two sheets 12 mm plasterboard on 90 mm metal studs, with absorbent quilt in air space — 46 dB

Figure 19

Mass-air resonance

Floating floors and dry-lined constructions are prone to a dip in performance caused by the MASS-AIR RESONANCE. This is the frequency at which the panel over the springy air pocket prefers to vibrate. To avoid an adverse dip in performance, the resonance must occur outside the frequency range of interest, usually well below 100 Hz. The resonant frequency is controlled by the panel mass and cavity width (see Figure 20 and 'Mass-air resonance' in box).

Control of flanking transmission

The sound insulation between two spaces depends not only on the sound transmitted through the intervening wall or floor, but also on FLANKING TRANSMISSION. This is the sound which travels along any elements common to both rooms (see Figure 21).

If flanking constructions are not correctly specified, flanking transmission can equal or even exceed direct transmission, and degrade the overall result. Flanking transmission is a complex matter. Not only are the construction and size of the flanking element significant but so is any interaction between it and the wall or floor. Specific guidance will be given for each construction type given in this manual.

The specification of a sound-insulating wall or floor is incomplete if not accompanied by an appropriate flanking construction specification.

Quality of detailing and workmanship

A wall or floor will give its expected sound insulation only if it and the surrounding construction are built without faults. Some possible faults are listed:

● Areas of low sound insulation; for example gaps or holes in the wall or floor (even those covered by plasterboard), areas of low mass.

● Mechanical bridging of isolation; for example, unwanted rigid connections such as nails through floating floors or the selection of incorrect ties in cavity walls.

● Acoustic coupling between the leaves of a double-leaf partition; for example, omission of a specified absorbent quilt or specification of a cavity which is too narrow.

● Excessive flanking transmission; for example, incorrect specification or incorrect bonding between the wall or floor and the flanking construction.

Absorption

It is important not to confuse the terms INSULATION and ABSORPTION. An absorber such as a fibrous quilt will not itself provide much sound INSULATION because it has low MASS and is permeable to air. Consequently, it is not a suitable material to fill up an aperture through a masonry partition if optimum sound insulation is required. Nor will it significantly improve the sound insulation of a domestic wall or floor if it is attached to one or both sides. However, absorbers can contribute to good sound INSULATION if installed in the cavity between leaves in double-leaf partitions, where they improve the acoustic ISOLATION between the two leaves.

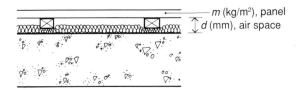

Figure 20

Mass-air resonance
(See Figure 20)

$$\text{Resonant frequency}, f_{res} = \frac{1900}{\sqrt{m \times d}} \text{Hz}$$

Examples

Construction	Air gap d (mm)	Resonant frequency (Hz)	$m \times d$
19 mm chipboard	50	72	700
22 mm floorboards	50	85	500
50 mm screed	6	72	690
19 mm plasterboard	25	87	365
19 mm chipboard	6	207	84

Rule $m \times d > 500$
The surface density, m (kg/m²), multiplied by the air gap, d (mm), should always exceed 500.

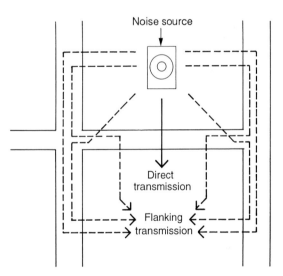

Figure 21

How much sound insulation is needed?

Separating walls and floors

The Building Regulations 1991 for England and Wales, the Building Regulations (Northern Ireland) 1990 and the Building Standards (Scotland) Regulations 1990 contain sound-insulation requirements for separating walls and floors in new-build dwellings and flat conversions. In all cases, the requirements can be met in two ways:

● by adopting an approved or deemed-to-satisfy form of construction, given to support the Regulations, or

● by demonstrating that a given numerical performance standard will be, or has been, achieved.

The requirements are summarised, country by country, on pages 18 and 19. Further details of the walls and floors covered by these Regulations are given on pages 26 and 27. While there are minor differences in the guidance for Northern Ireland, Scotland and England and Wales, revision is likely to remove them.

Approved or deemed-to-satisfy constructions

The constructions which support the Regulations or their accompanying documentation are reproduced in Part C. When using them there is no need to demonstrate that a given numerical performance will be or has been achieved.

Numerical performance

Where the construction adopted is not approved or deemed-to-satisfy the Regulations, the requirements are that the construction should be shown to be appropriate by test or precedent. In England and Wales and Northern Ireland, tests may be carried out on similar examples before the proposed wall or floor is constructed. In Scotland, post-construction testing may have to be carried out if it cannot be shown that the proposed construction will meet the requirements. Further details are given on pages 18 and 19.

Field and laboratory measurements should be carried out in accordance with the relevant part of British Standard BS 2750 at the 16 one-third octave band frequencies between 100 Hz and 3150 Hz. The results are then expressed as a single-figure weighted value according to the method given in British Standard BS 5821, as illustrated in the box. The numerical requirements are recorded, country by country, on pages 18 and 19.

Terminology (after BS 5821: 1984)

Airborne
Field tests:

Standardised level difference	D_{nT}
Weighted standardised level difference	$D_{nT,w}$*

Laboratory tests:

Sound reduction index	R
Weighted sound reduction index	R_w*

Impact
Field tests:

Standardised impact sound-pressure level	L'_{nT}
Weighted standardised impact sound-pressure level	$L'_{nT,w}$*

Laboratory tests:

Normalised impact sound-pressure level	L_n
Weighted normalised impact sound-pressure level	$L_{n,w}$*

* Terminology used in the various regulations

Other walls and partitions

General performance standards for walls and partitions which are not covered by building regulations are given on pages 26 to 29.

Airborne sound insulation
Example

Weighted standardised level difference, $D_{nT,w}$

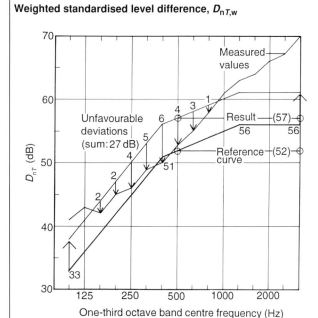

Figure 22

Procedure for obtaining $D_{nT,w}$

1 Plot the measured ONE-THIRD OCTAVE BAND D_{nT} values and the reference curve on graph paper, if possible to the nearest 0.1 dB.

2 Move the reference curve towards the measured curve in 1 dB steps.

3 Identify the frequencies where there is an unfavourable deviation. An unfavourable deviation occurs at a frequency where the measured value is less than the reference value.

4 Sum all the unfavourable deviations and divide by 16.

5 Repeat steps 2 to 4 until the figure obtained is as close as possible to 2.0 without exceeding it.

6 Read the value of the reference curve at 500 Hz. This is the $D_{nT,w}$ value in dB.

7 If the unfavourable deviation at any one frequency exceeds 8 dB, record it and the frequency at which it occurred.

Impact sound insulation
Example

Weighted standardised impact sound-pressure level, $L'_{nT,w}$

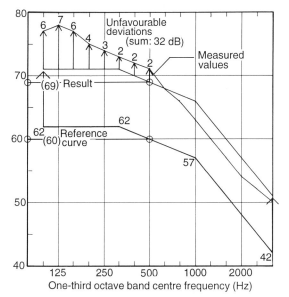

Figure 23

Procedure for obtaining $L'_{nT,w}$

1   Plot the measured ONE-THIRD OCTAVE BAND L'_{nT} values and the reference curve on graph paper, if possible, to the nearest 0.1 dB.

2   Move the reference curve towards the measured curve in 1 dB steps.

3   Identify the frequencies where there is an unfavourable deviation. An unfavourable deviation occurs at a frequency where the measured value is more than the reference value.

4   Sum all the unfavourable deviations and divide by 16.

5   Repeat steps 2 to 4 until the figure obtained is as close as possible to 2.0 without exceeding it.

6   Read the value of the reference curve at 500 Hz. This is the $L'_{nT,w}$ value in dB.

7   If the unfavourable deviation at any one frequency exceeds 8 dB, record it and the frequency at which it occurred.

Outside-to-inside sound insulation
External noise levels

The AIRBORNE SOUND INSULATION between the outside and the inside of a building depends on the sound-insulation performance of the individual elements in the building envelope. Under the Noise Insulation Regulations and airport authority grant schemes, secondary glazing, supplementary ventilation and, where necessary, double doors and improved roof insulation have been provided in exposed properties (see pages 10 and 11). For road traffic noise, the difference between external and internal levels in dB(A) can typically be improved to 34 (see page 30). It is difficult to achieve greater sound insulation than approximately 40 dB(A) in practice. Consequently, planners do not grant permission for housing development where the noise exposure is severe. Planning criteria for external noise levels are given on page 20. These should be consulted when assessing possible sites for housing.

Internal noise levels arising from external sources

Where sound-insulation measures are necessary, the degree of sound insulation required can be assessed if the external noise level is known and an appropriate design internal noise level is selected.

● The external noise level can be obtained by measuring it on site or by calculation (see page 10).

● An appropriate internal noise level can be selected by referring to the criteria for internal noise given on page 20.

● Methods are given on pages 30 and 31 for the design of the building envelope to achieve the necessary sound insulation.

Background noise

In assessing the effects of noise, it is sometimes appropriate to compare the level of an intrusive noise with the prevailing background noise conditions. This follows the simple principle that the higher the level of the intrusive noise relative to the background noise levels, the more clearly audible it becomes and the more it will be likely to bother residents. Though not generally used in the assessment of transportation noise sources, this principle is used in the assessment of industrial noise (see page 11) and amplified music.

In dwellings where the background noise level is low, noise from reasonable activity in neighbouring dwellings may become prominent, even where the separating elements comply with the Regulations. In these circumstances, it may be necessary either to increase the sound insulation of separating elements above regulation minima or to take steps to ensure that the background noise level is not too low. This should be taken into consideration when selecting appropriate noise criteria for building services (see below).

Internal noise levels arising from building services

Building services noise should be controlled to a reasonable level to avoid disturbance. The noise-control measures which the designer can take are described briefly on pages 32 and 33. The designer should also ensure that an appropriate steady background noise level is specified, by referring to the criteria on page 20 for internal noise levels resulting from mechanical services. The equipment specification should state that the selected level should not be exceeded when the equipment is in normal use, and that the sound should contain no distinguishable tonal or impulsive characteristics.

England and Wales

Part E of the Building Regulations 1991 contains sound-insulation requirements for separating walls and separating floors and stairs. Stated briefly, they are:

- separating walls shall resist the transmission of AIRBORNE SOUND, and

- separating floors and stairs shall resist the transmission of AIRBORNE and IMPACT SOUND.

The requirements apply both to new-build and to conversion dwellings. See page 26 for further details of the walls and floors governed by the Building Regulations.

Approved Document E: 1992 edition

There is no obligation to adopt any particular solution contained in the Approved Document if the designer prefers to meet the relevant requirement in some other way. However, should a contravention of a requirement then be alleged, if the designer has followed the guidance given in Approved Document E1/2/3 that will be evidence tending to show that he has complied with the Regulations. If he has not followed the guidance, he must show by other means that he has satisfied the requirements.

New-build construction

The Approved Document offers two ways in which the functional requirements can be met:

- By using one of the more widely used constructions which have been found to resist sound in practice. (These are reproduced in Part C .) When using them, there is no need to demonstrate that a given numerical performance will be achieved.

- By using a construction which is similar to one that has already been built in the field or in a test chamber and which has been shown to meet a given numerical performance standard.

New-build field numerical performance standards

Tests should be carried out in accordance with BS 2750 between at least 4 pairs of rooms, where possible (see Table 4). Where 8 pairs have been tested, a relaxation of 1 dB is given. Where it has been possible to test only one pair of rooms, the acceptable mean value for four measurements should be achieved.

Table 4

	Pairs of rooms tested	Acceptable mean value	Individual value
Airborne		(dB $D_{nT,w}$ (min))	(dB $D_{nT,w}$ (min))
Wall	Up to 4	53	49
	8 or more	52	49
Floor	Up to 4	52	48
	8 or more	51	48
Impact		(dB $L'_{nT,w}$(max))	(dB $L'_{nT,w}$(max))
Floor	Up to 4	61	65
	8 or more	62	65

When repeating an existing construction, the following features in the proposed building should be similar (but not necessarily identical) to those in the tested building:

- construction of the separating wall or floor, provided that the mass per square metre is not reduced;

- construction of other walls and floors adjacent to the separating wall or floor;

- arrangement of windows and doors in an external wall adjacent to the separating wall or floor when the external wall has a masonry inner leaf; and

- the extent of any step or stagger. (An increased or additional step or stagger is beneficial. See page 25.)

The following differences are allowed:

- construction of outer leaf of an external cavity masonry wall;

- construction of inner leaf of an external masonry cavity wall, provided that the construction is of the same general type and that the mass of the inner leaf is not reduced;

- material and thickness of floating layer of a separating floor with a concrete base (see pages 56 and 57); and

- construction of timber floor where it is not a separating floor.

New-build test chamber numerical performance standards

Section 4 of Approved Document E describes a method by which separating walls and their associated flanking constructions can be evaluated in a test chamber. Details of the approved test chamber construction can be obtained from the Building Research Establishment. A two-storey separating wall and its associated flanking construction are built in the chamber and one set of airborne test results is obtained at each level. The result is expressed as the modified weighted standardised level difference in dB. To obtain this value, the standardised level difference, $D_{nT,w}$ (see pages 12 and 16) is adjusted to allow for the length of the test chamber. In order to meet the functional requirements for walls, it is necessary for the tested wall to achieve a result of at least 55 dB.

Flat conversions

Where an existing wall, floor or stair is to become a separating element between dwellings, it is necessary either to show that it already meets the requirements or to adopt a treatment which will bring it up to standard. A construction will meet the requirements if either of the following can be demonstrated:

- the construction is generally similar to one of the Approved Document new-build constructions (for example within 15% of the mass of a construction in Part C of this manual), or

- the construction is shown to meet the field test performance standards given in Table 5.

There are two ways to meet the requirements if it cannot be demonstrated that the existing construction already meets them:

- Use one of the constructions reproduced in Part C, pages 64 to 71. When using them, there is no need to demonstrate that a given numerical performance will be achieved.

- Repeat an alternative treatment which has been built and tested in a building or a laboratory and which meets the numerical performance standards given in Table 5.

Table 5

	Minimum number of examples		Individual values	Units
Airborne			dB, minimum	
Wall	Field	2	49	$D_{nT,w}$
	Lab	1	53	R_w
Floor	Field	2	48	$D_{nT,w}$
	Lab	1	52	R_w
Impact			dB, maximum	
Floor	Field	2	65	$L'_{nT,w}$
	Lab	1	65	$L_{n,w}$

The treatment must be tested in conjunction with a representative base wall or floor. Typical examples are given on pages 64 (base wall) and 68 (base floor).

Northern Ireland

Part G of the Building Regulations (Northern Ireland) 1990 requires that separating walls provide adequate resistance to airborne sound transmission and that separating floors provide adequate resistance to airborne and impact sound transmission. See page 26 for further details on walls and floors.

Technical booklet G: 1990

This booklet contains deemed-to-satisfy provisions for new-build separating walls and floors. There are two deemed-to-satisfy methods.

- The acceptable construction method (section 1)
  The requirements can be satisfied by using one of the more widely used constructions which have been found to provide adequate sound insulation in practice. These are reproduced in Part C of this manual. When using them, there is no need to demonstrate that a given numerical performance will be achieved.

- The similar construction method (section 2)
  This method allows the repetition of a construction which has been built elsewhere, tested and shown to achieve given numerical performance standards (see Table 6).

New-build numerical performance standards

Tests should be carried out in accordance with BS 2750 between at least 4 pairs of rooms, or as close to 4 as possible (see Table 6). Where it has been possible to test only one pair of rooms, the acceptable mean value should be achieved.

Table 6

	Acceptable mean value	Individual value
Airborne	(dB $D_{nT,w}$ (min))	(dB $D_{nT,w}$ (min))
Wall	53	49
Floor	52	48
Impact	(dB $L'_{nT,w}$ (max))	(dB $L'_{nT,w}$ (max))
Floor	61	65

When using this method it is necessary to provide evidence that the tested construction achieved the sound-insulation values and that the proposed construction is essentially similar. The conditions on the use of a similar construction are much the same as those applied in England and Wales (see page 18).

Flat conversions

It is anticipated that the Regulations will be extended to cover flat conversions in early 1994. Technical booklet G1 will contain deemed-to-satisfy provisions for sound insulation where an existing wall or floor becomes a separating wall or a separating floor. The requirements are expected to be broadly as described for England and Wales, though not all the proposed floor treatments will be adopted and no treatment will be given for stairs (see pages 68 to 70).

Future revision

There are minor differences between Technical booklet G and Approved Document E for England and Wales. Future revision is likely to remove these.

Scotland

Part H of the Building Standards (Scotland) Regulations 1990 requires that separating walls provide adequate resistance to airborne sound transmission and that separating floors provide adequate resistance to airborne and impact sound transmission. See page 26 for further details on walls and floors.

Provisions deemed to satisfy the standards

The requirements contain deemed-to-satisfy provisions for separating walls and floors. There are two ways to meet the requirements.

- By using one of the more widely used constructions which have been found to provide adequate sound insulation in practice. These are reproduced in Part C of this manual. When using them, there is no need to demonstrate that a given numerical performance will be achieved.

- By adopting an alternative specification and carrying out sound-insulation tests after construction to demonstrate that it has met a given numerical performance standard (see Table 7).

Numerical performance standards

Tests should be carried out, in accordance with BS 2750, between at least 4 pairs of rooms, or as close to 4 as possible (see Table 7). Where it has been possible to test only one pair of rooms, the acceptable mean value should be achieved.

Table 7

	Acceptable mean value	Individual value
Airborne	(dB $D_{nT,w}$ (min))	(dB $D_{nT,w}$ (min))
Wall	53	49
Floor	52	48
Impact	(dB $L'_{nT,w}$ (max))	(dB $L'_{nT,w}$ (max))
Floor	61	65

Flat conversions

Flat conversions in Scotland are subject to the same requirements as new-build dwellings.

Future revision

There are minor differences between Part H of the Building Standards (Scotland) Regulations and Approved Document E for England and Wales. Future revision is likely to remove these differences.

Summary of noise control criteria

External noise levels

For new housing, individual planning authorities set their own standards, with due regard to guidance issued by the DOE. At the time of writing, DOE Circular 10/73, Planning and noise, is still current but draft Planning Policy Guidance (PPG) has been issued for comment and may be in use. Criteria from both 10/73 and the draft PPG are summarised here. As the former is likely to be superseded in the near future and the latter is subject to change, the reader should establish the latest information by contacting the appropriate local authority.

Circular 10/73

External noise criteria are given for road traffic, aircraft and industrial premises. The units used vary according to source type, and some are now obsolete. In most cases, free-field values are quoted (see page 9) but facade levels are used in the case of road traffic noise. To simplify the guidance given in Circular 10/73, the values are summarised in Table 8 below in terms of the estimated nearest equivalent $L_{Aeq,T}$ free-field values. For daytime levels, T has been taken as 16 hours (07.00 h to 23.00 h), and for night-time values T has been taken as 8 hours (23.00 h to 07.00 h).

Table 8

Source	Time	$L_{Aeq,T}$	Comment
Roads	Day	>65	There should be a strong presumption against permitting residential development
Aircraft	Day	>72	Refuse planning permission for dwellings
	Day	60 to 72	No major new developments. Infilling only with appropriate sound insulation
	Day	57 to 60	Permission not to be refused on noise grounds alone
Industrial	Day	>72	Scarcely ever justifiable to allow new developments which subject residential developments to these levels
	Night	>62	

Internal noise levels arising from external sources

Where sound-insulation measures are necessary, the degree of sound insulation required can be assessed if the external noise level is known and an appropriate design internal noise level is selected. The external noise level can be measured on site or predicted (see page 10). A design internal noise level can be selected using the criteria given in Table 10, based on British Standard BS 8233 and BRE Digest 266. Where good standards are required, 5 dB should be subtracted from the values.

Pages 30 and 31 contain methods for the design of the building envelope to achieve the necessary sound insulation.

Internal noise levels arising from mechanical services

Building services noise should be controlled to a reasonable level to avoid disturbance. The requirements should appear in any specification for equipment. An appropriate steady background noise level can be selected by reference to criteria given in Table 11, or by referring to CIBSE Guide A1 (see page 32). The specification should state that the selected level should not be exceeded when the equipment is in normal use and that the sound should contain no distinguishable tonal or impulsive characteristics.

Methods for controlling noise in building services are given on page 32.

The draft PPG

The DOE PPG, Planning and noise, defines four noise exposure categories (see Table 9).

Noise exposure category A

For proposals in this category, noise need not be considered as a determining factor in granting planning permission, although the noise level at the high end of the category should not be regarded as desirable.

Noise exposure category B

For proposals in this category, authorities should increasingly take noise into account when determining planning applications, and should require noise control measures.

Noise exposure category C

For proposals in this category, there should be a strong presumption against granting planning permission. Where it is considered that permission should be given, for example because there are no alternative sites available, conditions should be imposed to ensure adequate insulation against external noise.

Noise exposure category D

Planning permission should normally be refused.

Table 9

		Noise exposure category ($L_{Aeq,T}$)			
Source	Time	A	B	C	D
Road traffic	Day	<55	55 to 63	63 to 72	>72
Air traffic	Day	<57	57 to 66	66 to 72	>72
Rail traffic	Day	<55	55 to 65	65 to 74	>74
Mixed sources	Day	<55	55 to 63	63 to 72	>72
All sources	Night	<42	42 to 57	57 to 66	>66

Notes

For the day period, T is 16 hours; 07.00 h to 23.00 h.
For the night period, T is 8 hours; 23.00 h to 07.00 h.

The values refer to noise levels measured at 1.2 m to 1.5 m above the ground and at least 10 m away from any buildings. Levels measured 1 m from a facade should be assumed to be 3 dB(A) higher.

Where industrial noise dominates, an individual assessment should be made using BS 4142 (see page 11).

Table 10

Room classification	Suggested design background noise level arising from external noise sources (dB $L_{Aeq,T}$)		
Sensitive rooms			
Bedroom	<35	8 h	(23.00 h to 07.00 h)
Living room	<40	16 h	(07.00 h to 23.00 h)
Dining room	<40	16 h	(07.00 h to 23.00 h)
Less sensitive areas	<50	16 h	(07.00 h to 23.00 h)
Kitchen, bathroom, utility room, WC, internal and communal circulation areas			

Table 11

Room classification	Suggested design background noise level arising from mechanical services (dB(A))
Sensitive rooms:	
Bedroom	<30
Living room	<35
Dining room	<35
Less sensitive areas:	<45
Kitchen, bathroom, utility room, WC, internal and communal circulation areas	

Part B
Scheme design
Planning to control external noise

Noise control measures fall into three categories.

Control of noise at source

The types of external noise source described in Part A of this manual originate outside the boundaries of the site. The architect cannot usually control them at source, but can monitor the noise reaching the site, compare it with published criteria and then decide what sound control to include in the design.

Control of noise on the transmission path

Many factors influence the level of sound reaching the occupants of a dwelling. Those under the architect's control are:

● location of the building on site,

● screening of the site,

● internal planning of the building, and

● building form and orientation.

Control of noise at the receiver

The final line of defence against external noise is the building envelope, but there are two major drawbacks in relying on it for noise control.

● It does not shield the site from noise, so public areas and private gardens remain noisy.

● To achieve maximum sound insulation from the building envelope, the windows must be closed. Some form of mechanical ventilation will therefore be necessary.

Whenever possible, external noise should be controlled by site and building planning. Sound insulation of the building envelope should be used only as a last resort.

Planning to control internal noise

Internal noise control falls into two categories.

Internal planning

Dwellings should be planned to ensure that adjacent rooms are compatible in terms of noise sensitivity and noise production. Where good room-to-room sound insulation is required, the area of the intervening partition should be kept to a minimum and flanking paths eliminated where possible.

Sound insulation

Simply meeting the minimum standards of airborne and impact sound insulation associated with building regulations will not eliminate disturbance if adjacent rooms are incompatible.

Internal noise should be controlled by good internal planning combined with appropriate standards of airborne and impact sound insulation.

Cross-references	Page
External noise sources	10 and 11
Suggested criteria for external noise	20
Location of the building on site	22
Screening of the site	22 to 24
Internal planning of the building	23 and 24
Building form and orientation	22 to 24
Facade insulation	30 and 31
Constructional details	72 to 75

Cross-references	Page
Planning to control internal noise	25
Constructional principles	26 and 27
Constructional details	35 to 71

Control of noise on the transmission path

Location of the building on site

On a clear site, the main factors which control the noise level at any point are:

- distance between source and receiver,
- whether the ground cover is hard or soft, and
- height of the receiver.

Distance attenuation is greatest where there is soft-ground cover (such as grassland) and the receiver is near the ground. Therefore, low-rise housing can be built closer to a noise source than high-rise (see Figure 24).

Screening the site

Barriers or screens can reduce noise levels on site. They may take the form of:

- an existing feature (for example a cutting or elevated road),
- a purpose-designed feature (for example a solid boundary fence or an earth mound),
- a purpose-designed feature of the building (for example a courtyard), or
- a purpose-designed building (for example a barrier block).

For road traffic noise, the effect of a proposed barrier should be assessed using *Calculation of road traffic noise* (see page 10), which gives the attenuation in dB(A) . For the effect of a continuous barrier on other noise sources, a procedure is described in the box opposite.

The main points to bear in mind when designing barriers follow.

- They are most effective when located close to source or receiver.
- They protect low-rise housing better than high-rise.
- Generally, the taller the barrier the better but, as the rate of improvement with height diminishes progressively, there will come a point where increasing the height becomes relatively ineffective.
- They are particularly effective when the site slopes away from the source.
- They should usually extend well beyond the ends of the site to be protected.

Design of a purpose-built barrier

Sound transmitted through a barrier should be negligible compared with sound passing over or around it. The barrier material should be imperforate, but the construction need not be massive (see Table 12). In practice, appearance and stability tend to influence barrier height and construction. As a result, barriers are seldom more than 3 m high.

Table 12

Potential barrier attenuation (dB(A))	Required mass per unit area of unperforated cladding (kg/m²)
Up to 13	5
13 to 17	10

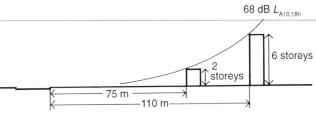

68 dB $L_{A10,18h}$

6 storeys

2 storeys

75 m

110 m

Motorway carrying 50 000 vehicles (20% heavy) per 18 h day
Average speed 80 km/h

Figure 24

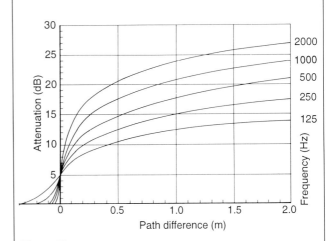

Source

Receiver

Barrier

Path difference = a + b − c

Figure 25

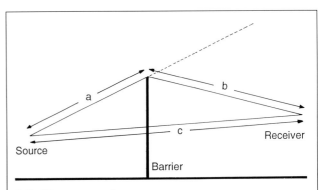

Attenuation (dB)

Frequency (Hz)

Path difference (m)

2000
1000
500
250
125

Figure 26

Procedure for assessing barrier attenuation
Measure or compute the sound level at the unscreened receiving point in octave bands.

Locate the proposed barrier on a section through the site and measure off the path difference as shown in Figure 25.

Using Figure 26, read off the attenuation at each frequency and subtract from the unscreened levels.

Using the procedures given on page 7, A-weight the result and sum the octave band levels to obtain a single-figure value in dB(A).

Barrier blocks

A barrier block is a building which itself forms a noise barrier. It consists of a continuous linear building situated close to and parallel to the noise source. As it will usually be taller than a purpose-built barrier, it is more effective and often able to reduce noise levels over the remainder of the site. But there are drawbacks:

- This is a high-density solution which may conflict with the density proposed for the site in the design brief.
- The ends of the block must be returned if the edges of the site are to be protected.
- Internal planning options are restricted if sensitive rooms are to face away from the noise source.
- Rooms on the noisy side of the building may need heavily insulated windows and mechanical ventilation.
- Consideration should be given to the orientation of the block and the need for sunlight.

Figure 27 shows a scheme where a barrier block screens low-rise housing.

Figure 27

Internal planning of a barrier block

Sensitive rooms should be protected from external noise by good room planning. Less noise-sensitive areas should be interposed between sensitive rooms and the noise source. Table 13 classifies rooms, with suggested design levels of background noise arising from external sources. Figure 28 shows examples of good internal planning against road-traffic noise in a barrier block.

Table 13

Room classification	Suggested design background noise level arising from external sources (dB $L_{Aeq,T}$)		
Sensitive rooms:			
Bedroom	< 35	8 h	(23.00 h to 07.00 h)
Living room	< 40	16 h	(07.00 h to 23.00 h)
Dining room	< 40	16 h	(07.00 h to 23.00 h)
Less sensitive areas:			
Kitchen, bathroom, utility room, WC, internal and communal circulation areas	< 50	16 h	(07.00 h to 23.00 h)

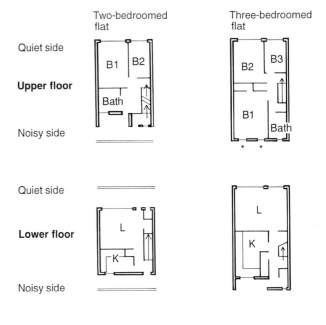

*Small, high insulation windows

Figure 28

Building form and orientation

Building groups

On an estate of low-rise dwellings, those closest to a noise source can give some protection to the remainder of the site. Figure 29 gives approximate values for the shielding given by two-storey houses built parallel to a main road. Even if the gaps between the houses amount to as much as 30% of the frontage length, 10 dB(A) attenuation can be provided over much of the site.

On sites where a small reduction in external noise is all that is required, it may prove sufficient to locate self-protecting dwellings, as described below, closest to the source to provide shielding to the remainder of the site.

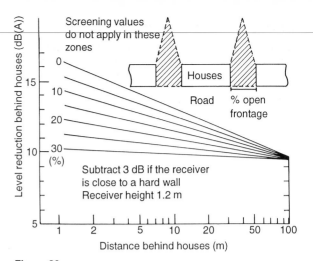

Figure 29

Self-protecting dwellings

Where barrier blocks are inappropriate, it still may be practicable to provide some protection against external noise using building form and orientation combined with good internal planning.

Staggered row of terraced housing

Figure 30 shows how a staggered row of low-rise terraced housing can be arranged to shield most windows from noise, though the dwellings are not parallel to the road.

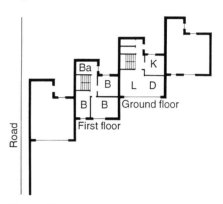

Figure 30

The courtyard house

The rooms overlook an internal courtyard but the external walls are windowless, providing visual and acoustic privacy (see Figure 31). The attenuation of external noise depends on the following design features:

● wall height — the higher the better,

● distance between the source and the external wall — the shorter the distance the better the barrier attenuation,

● courtyard depth — the deeper the better, and

● courtyard ground cover and openings in the courtyard walls — the more absorption the better.

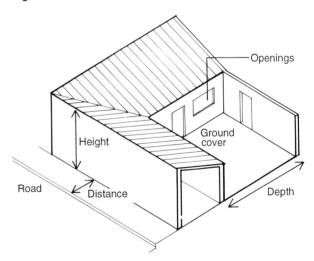

Figure 31

Balconies

A room facing a noise source can be given some protection by an external balcony (see Figure 32). This can reduce the received level by approximately 5 dB(A).

For road traffic noise, maximum protection is afforded when the building is close to the road. The balcony front and sides should be imperforate and as tall as possible. If stacked vertically, the underside of each balcony above should have a sound-absorbing finish, such as sprayed vermiculite.

A podium offers similar benefits to the lower floors of a high-rise building.

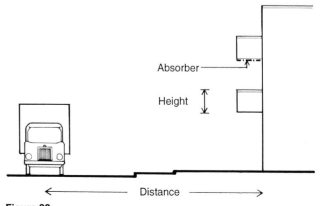

Figure 32

Planning to control internal noise

Adjacent rooms should be compatible in terms of noise production and sensitivity. 'Habitable' rooms, such as bedrooms, living rooms and dining rooms, are noise-sensitive. (Kitchens and bathrooms are not 'habitable' rooms.) Bedrooms are particularly sensitive to noise and should not be situated next to neighbours' living or dining rooms, kitchens, common circulation areas, bathrooms, lifts or other services areas.

Handing and stacking

Compatibility between rooms of adjacent dwellings can be ensured by HANDING and STACKING identical dwelling plans (see Figure 33). This may be impracticable or difficult to achieve in some situations, for example:

- ground-floor entrance hall/first-floor flat,
- top-floor flat/penthouse, or
- mixed occupancy flats.

Steps and staggers

A vertical displacement at the party wall between houses is known as a STEP. Steps occur on sloping sites. A horizontal displacement at the party wall is known as a STAGGER (see Figure 34). Each has the effect of reducing common wall areas and reducing flanking paths. Each improves sound insulation by approximately 3 dB. A step and a stagger together can give up to 6 dB improvement in sound insulation. A displacement must be at least 300 mm to be effective.

Where steps or staggers are a proposed design feature, consideration should be given to the following:

- Steps and staggers can have an adverse effect on planning, in terms of compatibility of adjacent rooms.
- Thermal insulation in the exposed part of staggered walls should be compatible with good sound-insulation detailing (as shown on pages 36 to 51).

Relative area of the separating wall

The sound transmitted through a separating wall is influenced by the common wall area. Figure 35 illustrates how the sound insulation between two rooms is affected by whether their short or their long dimensions coincide. In this example, halving the separating wall area results in a 3 dB (10 log 2) improvement in the level difference between the two rooms. Large separating wall areas, relative to other room dimensions should be avoided.

Quiet room

In some dwellings there may be a special requirement for a room designed either for quiet studying or to allow more than usual noise without disturbing other members of the household or neighbours. Where this is a design requirement, one room should be identified which can conveniently be insulated from other parts of the dwelling. Except in large houses, it will usually be necessary to provide partition walls and floors to an appropriate standard of sound insulation. Appropriate partition standards are given on page 27. Worked examples are given on pages 86, 104 and 105.

Buffer zones

Sensitive rooms can be protected from noisy areas by interposing less-sensitive rooms between them. For example, a kitchen may be located between a lift shaft and a living room, or an internal lobby between a common circulation area and a bedroom.

Service zones

A special zone should be allocated for mechanical and water services, and machine rooms should be kept away from sensitive rooms.

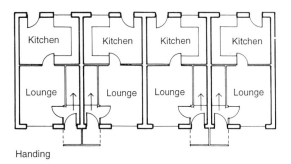

Handing

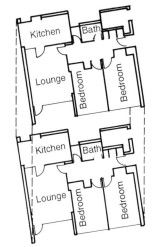

Stacking

Figure 33

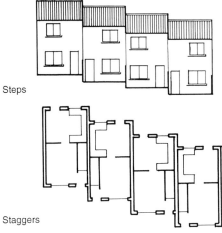

Steps

Staggers

Figure 34

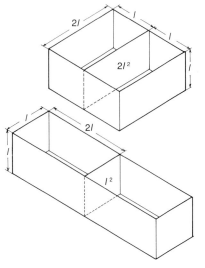

Figure 35

Constructional principles

New-build houses and low-rise flats

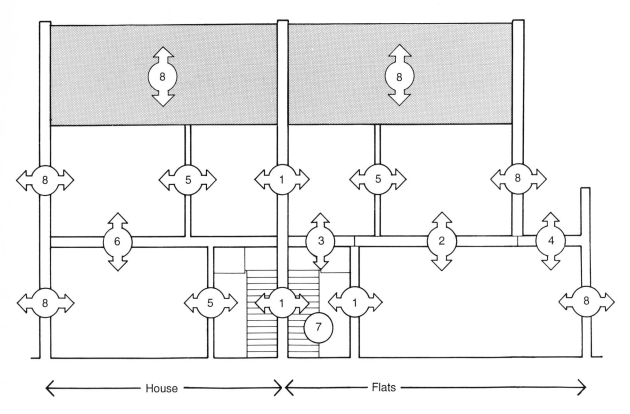

Figure 36 Section (See text for number references)

Walls and floors covered by the Building Regulations in England and Wales, Scotland and Northern Ireland

Sound insulation requirements apply only to dwellings. In all three regulatory geographical areas the requirements are based upon reasonable/adequate resistance to sound transmission. Applicants may propose their own solutions for approval by Building Control. However, most applicants choose to adopt one of the means of compliance described in documents that support the Regulations. In England and Wales these are in Approved Document E, in Scotland

in Technical Standard H, and in Northern Ireland in Technical Booklet G.

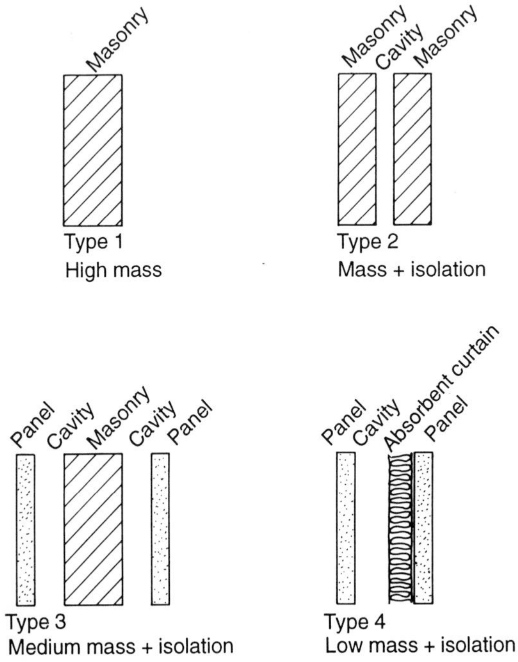

The main requirements of the Building Regulations are summarised in the following information which can be used as outline guidance at scheme inception stage. However, this information is not a substitute for the Regulations and supporting documentation. It is essential that the latest mandatory requirements and appropriate approving body be consulted at an early stage on specific projects, as there can be minor variations throughout the UK because of different timescales for implementation of the Regulations.

1 Separating walls (see Figure 36)
Between dwellings

Between a dwelling and another building

Between a dwelling and another part of the same building which is not part of the dwelling

Walls between dwellings and external areas are not covered by the Regulations.

Selection may be made from the four basic construction types shown in Figure 37. Further details of these walls and their associated flanking constructions are given on pages 36 to 51.

Alternatively, the requirements can be met by achieving specified numerical performance standards, which are given on pages 16 to 19.

Doors which separate dwellings from common circulation areas should have a minimum mass of 15 kg/m². A reduction in sound insulation should be avoided by providing a lobby. Both outer and inner lobby doors should be of minimum mass 15 kg/m². The outer door should be well sealed. Detailing of apertures, such as cat flaps, letter boxes and key holes, should be carefully considered (see page 73). The inner doors should be free from apertures, but need only to be hung to a normal good fit.

Type 1
High mass

Type 2
Mass + isolation

Type 3
Medium mass + isolation

Type 4
Low mass + isolation

Figure 37

2   Separating floors (see Figure 36)
Between dwellings

Between a dwelling and another part of the same building on the floor above which is not part of the dwelling

Selection may be made from the basic construction types shown in Figure 38. Further details of these floors and their associated flanking constructions are given on pages 52 to 63.

Alternatively, the requirements can be met by achieving specified numerical performance standards which are given on pages 16 to 19.

Floors between dwellings and external areas are excluded, except where an external walkway is above a dwelling (see 4 below).

3   Separating floors (see Figure 36)
Between a dwelling and another part of the same building on the floor below which is not part of the dwelling

Under the various Regulations, the impact requirement is waived for such floors as, for example, a floor separating a first-floor dwelling from a ground-floor lobby. Floor specifications should not be relaxed over a limited area unless it can be demonstrated that the airborne sound insulation will be unharmed, and the detailed flanking requirements are fulfilled around the floor area in question. In nearly all cases, floor specifications cannot be relaxed over a limited area without harming the airborne sound insulation.

4   Separating floors (see Figure 36)
Between a dwelling and an accessible area on the floor above

In these circumstances, the impact requirements must be met.

Walls and floors not covered by the Building Regulations

5   Partition wall (see Figure 36)
Suggested minimum standards:

- Quiet room partitions   48 dB $D_{nT,w}$

- Other domestic rooms   38 dB $D_{nT,w}$

Pages 39 to 51 give advice on the adjustment of the four basic wall specifications to obtain alternative standards of sound insulation. Page 65 gives practical guidance on the improvement of sound insulation in existing dwellings.

6   Partition floor (see Figure 36)
Suggested minimum standards:

- Above a quiet room   Airborne   46 dB $D_{nT,w}$
  (see page 25)   Impact: 68 dB $L'_{nT,w}$

- Below a quiet room   Airborne   46 dB $D_{nT,w}$

- Other domestic rooms   Airborne: 38 dB $D_{nT,w}$

Pages 44 to 49 give advice on the adjustment of the three basic floor specifications to obtain alternative standards of sound insulation. Floor treatments are given on pages 68 to 71.

7   Common lobbies and staircases (see Figure 36)
Residents tend to complain of impact noises such as doors banging and footsteps on stairs. These are not covered by the Building Regulations unless, in the case of stairs, the stair itself is part of a separating element. Wherever possible, carpet on underlay should be specified in lobbies and on stairs, to reduce impact sounds and provide useful absorption in otherwise hard areas.

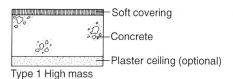
Type 1 High mass

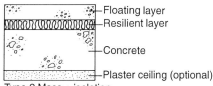
Type 2 Mass + isolation

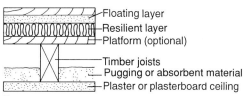
Type 3 Low mass + good isolation

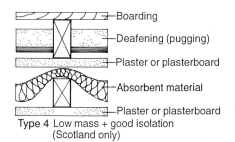
Type 4 Low mass + good isolation
(Scotland only)

Figure 38

Door banging may be alleviated by one of the following methods:

- Fixing compression seals which cushion the impact between door and frame (see page 73).

- Fitting appropriate closers which reduce the speed of impact.

Guidance on the sound insulation of lobby doors is given on page 26 and details are given on page 73.

8   Building envelope (see Figure 36)
See pages 30 and 31.

New-build medium- and high-rise flats
Additional walls and floors covered by Building Regulations

Listed here are situations which are likely to arise only in the design of medium- and high-rise dwellings.

Separating wall
> Between a habitable room or kitchen and a refuse chute

The guidance suggests a wall mass of at least 1320 kg/m², including any plaster finishes (see Figure 39).

Separating wall
> Between a non-habitable room other than a kitchen and a refuse chute

The guidance suggests a wall mass of at least 220 kg/m² (see Figure 39).

Separating wall
> Between a habitable room and a machinery room* or tank room

Walls should meet the airborne standards deemed-to-satisfy the Regulations (see a in Figure 40).

Dwelling
> Below a machinery room* or tank room

Floors should meet the airborne and impact standards deemed-to-satisfy the Regulations (see b in Figure 40).

Dwelling
> Above a machinery room* or tank room

Floors should meet the airborne standards deemed-to-satisfy the Regulations (see c in Figure 40).

* Whether this will be satisfactory in practice depends on the noise level of the proposed machinery and the background noise in the dwelling.

Quiet machinery should be selected, and it should be installed on suitable anti-vibration mounts. Lift shafts and motor-room walls which are also separating walls should have at least this degree of sound insulation.

Further information on lifts and other domestic services is given on pages 32 to 34.

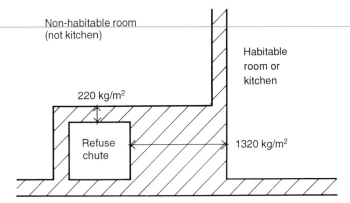

Figure 39

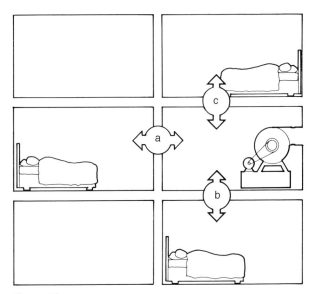

Figure 40

Flat conversions

In England and Wales, there are separate provisions for flat conversions. These are summarised below under the headings 'Separating walls', 'Separating floors' and 'Stairs'.

In Northern Ireland it is anticipated that the Regulations will be extended to cover flat conversions in early 1994. The requirements are expected to be broadly as described below for England and Wales. In Scotland, flat conversions are subject to the same sound-insulation requirements as new dwellings (see pages 17, 26 and 27).

Separating walls (England and Wales)

Where the existing wall is not similar to an Approved Document new-build separating wall, and cannot meet the numerical performance standards given on pages 18 and 19, an independent lining with absorbent material in the cavity should be adopted (see Figure 41). Further constructional details for separating walls and the associated flanking elements are given on pages 64 and 65.

An alternative treatment may be adopted if, in association with the existing wall, it can achieve the numerical performance standards given on pages 18 and 19.

Separating floors (England and Wales)

Where the existing floor is not similar to an Approved Document new-build separating floor, and cannot meet the numerical performance standards given on pages 18 and 19, floor treatment 1, 2 or 3 shown in Figure 42 should be adopted.

Floor treatments 4 and 5 should be used only when a strong case can be made for not using floor treatments 1, 2 and 3.

Further constructional details for separating floors and the associated flanking elements are given on pages 64 and 65.

An alternative treatment may be adopted if, in association with the existing floor, it can achieve the numerical performance standards given on pages 18 and 19.

Stairs (England and Wales)

Where a timber stair performs a separating function between dwellings, a soft covering should be applied to the treads and an independent ceiling with absorbent material should be installed below (see Figure 43). Further details are given on page 70.

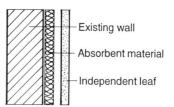

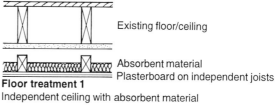

- Existing wall
- Absorbent material
- Independent leaf

Figure 41

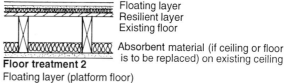

Existing floor/ceiling

Absorbent material
Plasterboard on independent joists
Floor treatment 1
Independent ceiling with absorbent material

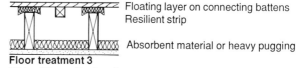

Floating layer
Resilient layer
Existing floor

Absorbent material (if ceiling or floor is to be replaced) on existing ceiling
Floor treatment 2
Floating layer (platform floor)

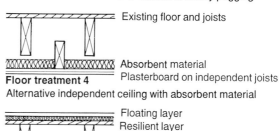

Floating layer on connecting battens
Resilient strip

Absorbent material or heavy pugging
Floor treatment 3
Ribbed floor with absorbent material or heavy pugging

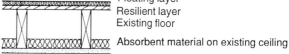

Existing floor and joists

Absorbent material
Plasterboard on independent joists
Floor treatment 4
Alternative independent ceiling with absorbent material

Floating layer
Resilient layer
Existing floor

Absorbent material on existing ceiling
Floor treatment 5
Alternative floating layer (platform floor)

Figure 42

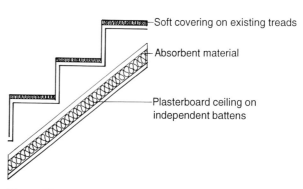

- Soft covering on existing treads
- Absorbent material
- Plasterboard ceiling on independent battens

Figure 43

Building envelope

The following information is required for accurate calculation of sound-insulation requirements of the building envelope (see box):

External noise level

Maximum allowable sound level in the room

Surface area of the relevant portion of the building envelope

Area of sound absorption in the room

Formula to determine facade sound insulation against road-traffic noise (see also page 13)

Level difference = $L_1 - L_2 = R - 10 \log S/A$ (dB)

where

L_1 = Sound level 2 m outside the facade (dB)

L_2 = Received sound level in the room (dB)

R = Sound reduction index (dB)

S = Surface area, room facade element (m^2)

A = Absorption in the room (m^2)

For housing design purposes, a more simple approach is proposed:

● The surface area and area of sound absorption can be ignored. In typically-furnished domestic rooms they have little effect on the final result.

● A typical external noise spectrum is adopted and the sound insulation of the building envelope described in terms of the difference between outside and inside levels in dB(A).

This outside-inside level difference, $R_{A(\text{traffic})}$, is based on the typical urban road-traffic noise spectrum illustrated in Figure 44. Representative values for building components are given in Table 14. (Values for free-flowing motorway traffic would be 1 dB to 3 dB higher because their spectra contain relatively less low-frequency noise). Table 14 also gives typical sound-insulation values for building envelope elements in terms of two other single-figure units commonly quoted by manufacturers and testing laboratories:

R_m The average sound reduction index taken over 16 one-third octave bands (100 Hz to 3150 Hz), and

R_w The weighted sound reduction index (see page 16).

The values in Table 14 are representative and will be adequate for most design situations. However, if possible, it is better to use test values for the actual form of construction under consideration.

Walls

Masonry walls have better sound insulation than the other elements in the building envelope. To achieve $R_{A(\text{traffic})}$ values of up to approximately 40 dB, it is necessary to consider the specification, detailing and construction of only the windows, doors, roof and lightweight cladding, and the method of ventilation.

Walls built of lightweight materials have lower sound insulation than masonry walls at low frequencies. As road-traffic noise peaks at low frequencies, the resulting $R_{A(\text{traffic})}$ can be low; as low as 25 dB in some cases. On noisy sites, lightweight cladding should be avoided unless specifically designed to provide adequate sound insulation.

Ventilation

Standards for background ventilation in habitable rooms are given in Approved Document F. Table 14 and Figure 45 give sound-reduction index values for a brick wall containing a unit ventilator meeting the requirements of the Noise Insulation Regulations (NIR) (see page 10), for a trickle vent in a single-glazed window (a trickle vent and thermal double glazing would be similar), and for secondary glazing with absorbent-lined reveals and staggered opening lights. Page 32 gives guidance on mechanical ventilation which may be necessary where higher standards are required.

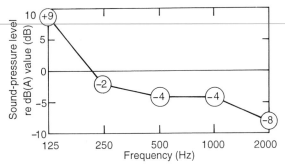

Figure 44

Table 14

Building envelope element	R_m	R_w	$R_{A(\text{traffic})}$	See Figure
Walls				
Plastered solid brickwork (480 kg/m^2)	50	54	47	45
As above, but NIR ventilator in wall	46	51	39	45
Plastered solid brickwork (260 kg/m^2)	45	48	41	45
112.5 mm brickwork/100 mm lightweight blockwork plastered one side, 50 mm cavity	50–53	–	45–50	
Timber-frame /112.5 mm brick facing	45–49	–	40–46	
Timber-frame /hung tiles facing	39–42	–	34–39	
Windows				
Values for fixed or well-sealed openable windows but without absorbent material on the window reveals				
Single-glazed, 4 mm	25	28	24	46
Single-glazed, 6 mm	27	29	26	46
Single-glazed, 12 mm	31	33	29	46
Double-glazed, 6/12/6	28	31	26	46
Double windows, 6/100/6	37	40	34	46
Double windows, 6/200/6	44	46	41	46
For absorbent on double window reveals, add: 3		3	1	46
Window ventilation				
Trickle vent, 435 mm × 16 mm slot:				
In single-glazed window	25	27	25	45
In secondary-glazed window	29	28	28	45
Double windows, 4/250/4,	22	26	15	45
area 2.4 m × 1.3 m staggered open leaves, (200 mm × 1300 mm open), absorbent-lined reveals				
Doors				
Hollow-core, no seals	15	17	15	47
Solid-core, 25 kg/m^2, no seals	20	21	19	47
Solid-core, 25 kg/m^2, well sealed	27	30	25	47
Solid-core, 40 kg/m^2, well sealed	33	36	31	47
Roofs				
Pitched, tiles on felt roof, 9 mm plasterboard ceiling	30	34	27	48
As above, 100 mm absorber on ceiling	39	43	31	48
Pitched, tiles on felt, wood-lath and plaster ceiling	43	47	37	48
As above, 100 mm absorber on ceiling	47	51	38	48
Flat, 100 mm concrete (230 kg/m^2)	49	52	45	48
Flat, timber-joist roof, asphalt on boarding, 12 mm plasterboard ceiling, thermal insulation	43	45	30	48

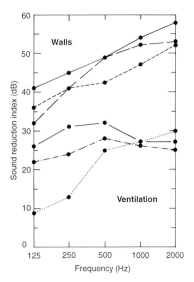

Walls

Ventilation

See Table 14 for key to graph

Figure 45

Windows

Typical sound reduction index values are given in Table 14. Octave band values are given in Figure 46. The main factors determining the sound insulation of windows are:

● The thicker the glass, the higher its sound insulation, subject to a CRITICAL FREQUENCY dip at high frequencies (see page 14)

● The wider the cavity between two panes, the higher the sound insulation, subject to a MASS-AIR RESONANCE (see page 15)

● The presence of absorbent materials on the window reveals improves sound insulation

● To achieve their optimum sound performance, windows must make an airtight seal with their frames and their frames must make an airtight seal with the opening. Openable windows must be carefully detailed to achieve a good seal in practice

Where the external window is double-glazed (for thermal insulation), secondary single glazing can still be used to improve sound insulation and the above considerations will still apply.

Ventilation requirements may conflict with both the thermal and the sound-insulation requirements of windows. For better sound insulation it may be necessary to have the ventilator separate from the window, or, where high standards are required, mechanical ventilation may be necessary. Representative values associated with various ventilation arrangements are given in Figure 45 and Table 14.

Doors

Sound-insulation values for four types of domestic door are given in Table 14 and Figure 47. The main factors determining the sound insulation of doors are:

● The sound insulation of any door will be maximised by fitting perimeter seals to make it airtight and avoiding apertures such as cat flaps. Closing flaps should be provided for letter boxes and key holes.

● The heavier the door, the higher its sound insulation.

● To maximise the sound insulation of the building envelope, lobbies should be provided.

Roofs

Pitched roofs

Table 14 and Figure 48 give sound-insulation values for four roof/ceiling combinations. The sound insulation of pitched roofs may be increased by:

● increasing the mass of the ceiling (or roof), or

● providing an absorbent material above the ceiling.

Flat roofs

Table 14 and Figure 48 give representative values of sound insulation for a concrete roof and a timber-joist roof. Their sound-insulation performance is similar to that of separating floors of similar construction.

Note

The $R_{A(traffic)}$ value may underestimate the resistance of roofs to aircraft noise by approximately 3 dB. Guidance is given in the example on page 98.

Composite insulation

The sound insulation provided by a wall containing an element such as a window depends on the following:

● the sound reduction index of the window,

● the sound reduction index of the wall, and

● the surface areas of wall and window.

Figure 49 combines all these factors in a chart, which gives a resultant $R_{A(traffic)}$ for a variety of configurations. It can be used to assess the composite sound insulation of any combination of materials, not only windows and walls.

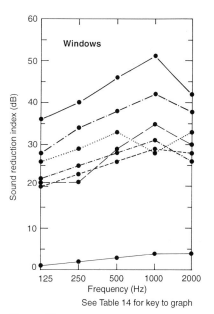

Figure 46

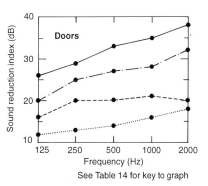

Figure 47

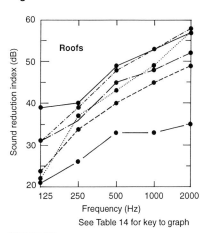

Figure 48

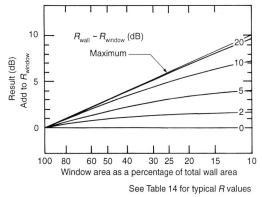

Figure 49

Building services

Introduction
Good planning solves many potential noise problems associated with domestic services.

The risk of noise problems can be limited at an early stage by selection of intrinsically quiet systems. Where appropriate, passive ventilation should be used, as long as it is compatible with the sound-insulation requirements of the building envelope (see pages 30 and 31).

The architect should be aware of the need to locate noisy items of equipment, such as fans, away from sensitive rooms and should ensure that appropriate noise requirements are included in equipment specifications and that these specifications are met.

Mechanical ventilation
Mechanical ventilation may be provided in one of two forms; room ventilator units, or a ducted air ventilation system. These are described below.

Room ventilator units
Noise-attenuated ventilator units have been specified extensively under the Noise Insulation Regulations (see page 10) and airport authority grant schemes (see page 11). They comprise:

- a variable speed air supply unit which is designed to be used in the facade of noise-exposed rooms, and

- a permanent air outlet vent,

or a single unit which combines both.

The Noise Insulation Regulations give detailed performance specifications which include the following:

Airflow:	At least two specified settings
Self noise:	High setting less than 40 dB(A) Low setting less than 35 dB(A)

Sound insulation against external noise sources:
See page 30, Figure 45, for details

Manufacturers must obtain an Agrément Certificate showing that their units comply with the above requirements before they can be specified under the Noise Insulation Regulations.

Ducted air ventilation systems
A brief guide to the control of noise from ducted air ventilation systems is given here. More detailed advice can be found in Guides A1 and B12 published by the Chartered Institution of Building Services Engineers (CIBSE).

The main elements of a ducted air ventilation system are the air-moving plant, which may or may not be located in a plantroom, and the ducts which distribute the air to and from rooms.

The design background noise level should be not more than 35 dB(A) in sensitive rooms and not more than 45 dB(A) in less sensitive rooms. (CIBSE Guide A1 gives criteria in an alternative form.) The specified level should not be exceeded when the equipment is in normal use and the sound should contain no distinguishable tonal or impulsive characteristics.

The main acoustic design factors are presented as a checklist in the box.

If inadequate space is provided or sizes of units are too small, noise problems are more likely to occur. For example, undersized fans make more noise; smaller cross-section ducts must carry higher air velocities which cause more noise, and silencers take up more space than the ducts themselves. Space and size requirements should be agreed with the services engineer at an early stage to avoid noise problems later.

How is building services noise transmitted?
Noise from domestic plant, machines and appliances can be transmitted in two ways:

Airborne sound transmission
Sound is transmitted to the building construction via the air.

Structureborne sound transmission
When the source is in direct contact with the building, vibration may be transmitted directly from the source into the building materials. The vibration can be transmitted through the building elements and reradiated as sound in other rooms. The worked examples on pages 107 to 109 give guidance on the identification and control of structureborne sound.

Ducted ventilation systems: noise control checklist

Plant location and design
- Keep noisy plant away from sensitive rooms.

- Provide adequate sound insulation in plantroom walls, floors, windows and doors.

- Ensure floor can support plant and vibration isolation inertia block loads.

Plant selection and installation
- Select a fan which will operate at its optimum efficiency.

- Create favourable air-flow conditions through the fan by avoiding abrupt changes in the direction or cross-section of ducts close to the fan.

- Mechanically isolate fans from the building structure.

- Install flexible connectors between fan and ductwork.

Duct design
- Avoid routing ducts through sensitive rooms.

- Seal around ducts where they pass through partitions, separating walls and floors.

- Minimise airflow-generated noise by designing simple layouts with radius bends, flared sections and smooth surfaces.

Silencers (and absorbent duct linings)
- In-duct silencers can be used to reduce fan noise, exterior noise and transmission between rooms via inter-communicating ducts ('cross-talk' transmission).

- Reduce fan noise by locating the silencer close to the fan.

- Where necessary to maintain partition sound insulation, locate cross-talk silencers where the duct passes through the partition.

Duct lagging
- Lag ducts to reduce noise transmission through the duct walls, either break-out or break-in.

Water services

The most serious noise problems in domestic water installations are likely to result from the following:

● Poor internal planning, for example locating a communal water storage tank above a bedroom (Figure 50) or routing a soil stack (which can be a source of rattling noise) through a habitable room

● High velocity water passing through valves and taps

● Pipes carrying mains pressure or mechanically-pumped water and which are rigidly fixed to lightweight constructions

● Movement of pipes, particularly heating and hot water pipes and plastic waste and soil pipes

CIBSE Guide B12 discusses noise control mechanisms and criteria for water systems in more detail.

Noise control measures

Precautions which should be taken at the design stage are listed below.

Water services should be zoned away from sensitive rooms. Where this is not possible, water should be supplied by gravity feed from a cistern, rather than using mains pressure.

Quiet equipment should be specified either by reference to any laboratory or field test data provided by manufacturers or by visiting and assessing equipment in use.

Incoming mains water pipes should be routed next to heavy walls and, where possible, grouped together and contained in masonry shafts. Where masonry shafts are not feasible, pipes should be wrapped in mineral or glass fibre and encased in a plasterboard shaft.

Potentially noisy pipes and those subject to thermal movement should have a resilient sleeve (see Figure 51).

Access panels in services shafts should be located away from sensitive rooms. To maximise the sound insulation, a heavy panel should be used with compression seals fitted to the stops to create an airtight seal.

To prevent a reduction in room-to-room sound insulation where pipes pass through partitions, a flexible but airtight seal should be provided. Pipes should be wrapped with a resilient material where they pass through floating floors to prevent bridging of the floor isolation.

WC cisterns should be located on heavy walls to minimise noise in the same dwelling. To minimise noise in neighbouring dwellings, WC cisterns should not be fixed to party walls or flanking walls, if possible.

The airborne sound insulation between WCs and sensitive rooms should be at least 38 dB $D_{nT,w}$.

Where a WC, bath, kitchen unit, etc must be installed on top of a floating floor, the fixings should not give rise to a rigid connection between the floating layer and the joists. In the case of baths, the loading should be spread to avoid excessive compression of the resilient layer.

Water pumps should be mechanically isolated from the building structure and flexible connectors should be installed between pump and pipework (see the worked example on page 108). Noise from services in airing cupboards can be reduced by fitting heavy, well-sealed doors, as long as structureborne transmission has been adequately controlled.

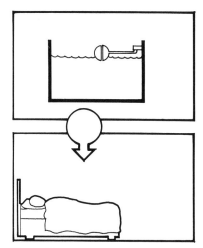

Figure 50

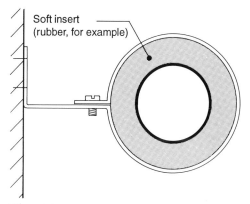

Soft insert
(rubber, for example)

Figure 51

Electrical services

The following precautions should be taken to prevent the electrical services from affecting room-to-room sound insulation.

Socket outlets should not be installed back-to-back on separating walls.

On lightweight separating walls and floors, any penetrations for electrical cables should be sealed and socket boxes enclosed to avoid air leaks (see worked example on page 83 for a suggested detail).

Electrical conduits should not be run between a floating floor and its base construction.

Lifts

The following precautions should be taken to control lift noise.

The lift shaft and motor room should be located away from sensitive rooms. Hydraulic lift motors, which are suitable for low- and medium-rise buildings, are normally situated in a basement where it is easier to control structureborne noise transmission.

A quiet lift should be selected by comparing manufacturers' noise data or by visiting and appraising an existing installation. The design noise level should be no more than 35 dB(A) in sensitive rooms and 45 dB(A) in less sensitive rooms.

The sound insulation between the lift motor room and neighbouring flats should be in accordance with the requirements of the Building Regulations.

A lift shaft of mass 200 kg/m^2 or more should be provided (unless it is next to a flat, in which case, the requirements of the Building Regulations must be met).

The lift motor should be mechanically isolated from the building structure.

Other sources of lift noise include switchgear, cage guide rollers and lift doors. These should be of quiet design and must be maintained regularly to ensure continued quiet operation. The client should be advised to make appropriate arrangements for regular maintenance.

Boilers

The architect should consider the following, in schemes which include a boiler house.

The boiler house should be sited away from sensitive rooms. A separate boiler house is the ideal. Boiler rooms in the same building as the dwellings will cause problems unless great care is taken to control airborne and structureborne noise.

The plant should be mechanically isolated from the building structure.

Where pipes pass through walls and floor slabs they should be sleeved in mineral fibre or similar material to prevent structureborne sound transmission.

If the boiler house must be naturally ventilated, ventilation louvres should be located well away from sensitive areas. If this is not possible, a noise expert should be consulted.

Adequate attenuation should be provided for noise which may be emitted from the boiler exhaust flue.

Domestic appliances

Washing machine

There is no substitute for good siting of the washing machine. Ideally, it should be on a solid ground floor. Effective mechanical isolation is not practicable as typical spin speeds fall in the low-frequency range between 10 Hz and 20 Hz. Vibration isolators would cause the machine to vibrate excessively unless a substantial concrete inertia block were installed.

The most troublesome location for a washing machine is on a lightweight upper floor where the resonant frequency of the joists may be near the machine spin frequency. As a result, vibration amplification can occur. To minimise the effects in practice, the washing machine should be positioned near the end of joists, where the deflection is least.

Other domestic appliances

Other domestic appliances are likely to pose lesser problems. They should be selected for quiet operation and should be installed so that they are not in rigid contact with separating walls or lightweight partitions.

Part C
Detailing

(Figures 52 and 53)

Part C gives the details of construction for good sound insulation. All the separating-wall and floor constructions which satisfy the current Building Regulations for England and Wales, Northern Ireland and Scotland appear here. Guidance is also given on the specification of walls and floors to achieve standards other than those required by the Regulations.

It should be stressed that sound insulation is only one design requirement, and that there may be cases where requirements for sound conflict with other requirements. A list of the other aspects of design which are the subject of building regulations follows. Due consideration should be given to each when adopting any detail for sound insulation.

    Structure
    Fire safety
    Site preparation and resistance to moisture
    Toxic substances
    Ventilation
    Hygiene
    Drainage and waste disposal
    Heat producing appliances
    Stairs, ramps and guards
    Conservation of fuel power
    Access and facilities for disable people
    Glazing: materials and protection
    Materials and workmanship

The building elements covered in Part C are listed opposite. Descriptions of each element are presented as follows:

Factors affecting performance
The relative importance of mass, isolation and flanking transmission is stated briefly.

Approved or deemed-to-satisfy constructions
All constructions which comply with the technical specifications given in the following documents are included:

    Approved Document E, 1992 Edition, The Building Regulations 1991

    Technical Booklets G, 1990 and G1, 1992, The Building Regulations (Northern Ireland) 1990

    Part H, 1990, of the Building Standards (Scotland) Regulations 1990

Although all the approved and deemed-to-satisfy constructions are reproduced in this manual, the designer should ensure that the design complies with the current source document.

Other separating element constructions
Summaries are given of other ways in which the various regulations can be met. Where available, examples of specific constructions are given. These constructions are neither 'approved' nor 'deemed-to-satisfy' but may meet the requirements in appropriate circumstances.

Elements of lower and higher performance
Guidance is given on how to adjust the specification of a separating element to achieve higher or lower standards of sound insulation.

Checklists
Two checklists are given, for use during design and during site inspections.

Information sources
Sources of further information. Their addresses may be found in Appendix F on page 128.
At the end of Part C, guidance is given on the detailing of building envelope components, windows, doors and roofs, for good sound insulation.

Construction details: separating walls

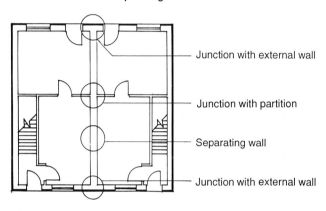

(a) Plan

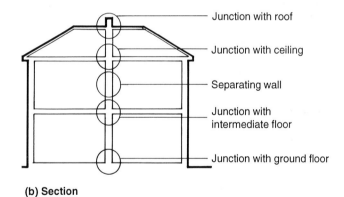

(b) Section

Figure 52

Construction details: separating floors

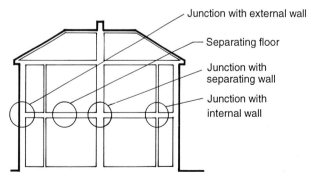

Figure 53

Solid masonry walls

Factors affecting performance

The AIRBORNE SOUND INSULATION of solid masonry walls is controlled mainly by MASS, QUALITY OF CONSTRUCTION and FLANKING TRANSMISSION.

England and Wales, Approved Document E

The constructions which appear under this heading comply with Approved Document E, 1992 Edition, The Building Regulations 1991. It is not necessary to demonstrate that these constructions will meet a given numerical performance standard. The mean performance figures are given only to assist the designer.

Construction A
Brickwork plastered both sides
(Figure 54(a))

Mass 375 kg/m² or more (including plaster)

Quality Bricks laid frog-up. Joints fully filled and sealed with mortar. Junctions with surrounding constructions sealed

Example 215 mm brickwork, brick density 1610 kg/m³
75 mm coursing
13 mm lightweight plaster

Construction B
Concrete blockwork plastered both sides
(Figure 54(b))

Mass 415 kg/m² or more (including plaster)

Quality Blocks extend across the full thickness of the wall. Joints fully filled with mortar. Junctions with surrounding constructions sealed

Example 215 mm blockwork, block density 1840 kg/m³
110 mm coursing
13 mm lightweight plaster

Construction C
Brickwork with plasterboard both sides
(Figure 54(c))

Mass 375 kg/m² or more (including plasterboard)

Quality Bricks laid frog-up. Joints fully filled and sealed with mortar. Junctions with surrounding constructions sealed, including those behind linings. Plasterboard joints sealed

Example 215 mm brickwork, brick density 1610 kg/m³
75 mm coursing
12.5 mm plasterboard

Construction D
Concrete blockwork with plasterboard both sides
(Figure 54(d))

Mass 415 kg/m² or more (including plasterboard)

Quality Blocks extend across the full thickness of the wall. Joints fully filled with mortar. Junctions with surrounding constructions sealed, including those behind linings. Plasterboard joints sealed

Example 215 mm blockwork, block density 1840 kg/m³
150 mm coursing
12.5 mm plasterboard

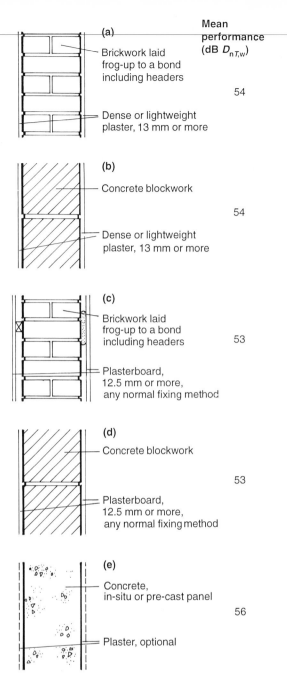

Mean performance (dB $D_{nT,w}$)

(a) Brickwork laid frog-up to a bond including headers — Dense or lightweight plaster, 13 mm or more — 54

(b) Concrete blockwork — Dense or lightweight plaster, 13 mm or more — 54

(c) Brickwork laid frog-up to a bond including headers — Plasterboard, 12.5 mm or more, any normal fixing method — 53

(d) Concrete blockwork — Plasterboard, 12.5 mm or more, any normal fixing method — 53

(e) Concrete, in-situ or pre-cast panel — Plaster, optional — 56

Figure 54

Construction E
Concrete, in-situ or large pre-cast panel
(minimum density 1500 kg/m³)
(Figure 54(e))

Mass 415 kg/m² or more (including plaster, if used)

Quality Joints between panels filled with mortar, and any honeycombing made good. Junctions with surrounding constructions sealed

Example 190 mm concrete, density 2200 kg/m³. No plaster

Note Where plaster or plasterboard finish is specified, wall lining laminates of plasterboard and mineral fibre may be used instead.

Control of flanking transmission

In section

(Figure 55(a))

The joint between the wall and the roof should be filled. A method used for fire stopping, such as mortar bedding or mineral fibre, should be used.

In the roof space, the wall mass may be reduced to not less than 150 kg/m² , but only if the ceiling mass is equivalent to at least 12.5 mm plasterboard and the joints are well sealed. If lightweight aggregate blocks of density less than 1200 kg/m³ are used, at least one side should be sealed using cement paint or plaster skim.

At the junction between separating wall and ground or intermediate floors, floor joists may be supported on hangers or built in. They may be built in only if good workmanship can be assured, and care should be taken to ensure that there are no airpaths through the wall. All gaps should be sealed. With concrete floor constructions, described on pages 52 to 59, either the wall or the floor may be carried through.

In plan

(Figure 55(b))

The outer leaf of a cavity wall may be of any construction.

Where a cavity wall has an inner leaf of masonry, or where the external wall is solid masonry, the mass of the flanking leaf should be at least 120 kg/m² unless the conditions below are met (see 'Openings in the external wall').

Either the masonry of the flanking wall should be bonded to that of the separating wall, or the masonry of the flanking wall should abut the separating wall and be tied to it with ties at no more than 300-mm centres vertically

Where the external wall has a cavity, the cavity should be sealed with a flexible closer (for example mineral wool).

Where a cavity wall has an inner leaf of timber construction:

● a tight butt-joint should be formed between the timber leaf and the separating wall,

● the timber leaf should be tied to the separating wall with ties at no more than 300-mm centres vertically, and

● the joints should be sealed with tape or caulking.

Openings in the external wall

If the flanking leaf is of masonry construction, its mass should be at least 120 kg/m² unless there are openings at least 1 m high within 700 mm of each side of the separating wall (see Figure 56). These openings are required on every storey.

(Openings close to the separating wall have the effect of restricting the flow of energy between the major portions of the flanking walls on either side of the separating wall. This reduces flanking transmission.)

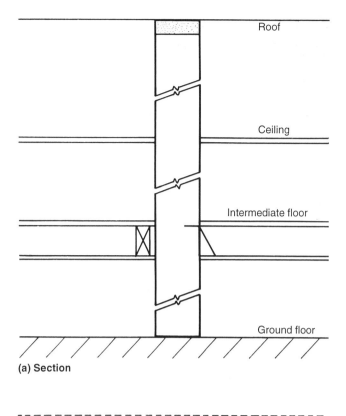

(a) Section

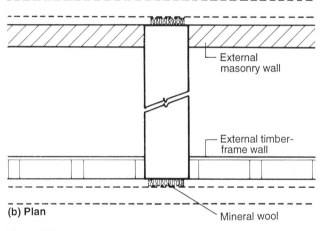

(b) Plan

Figure 55

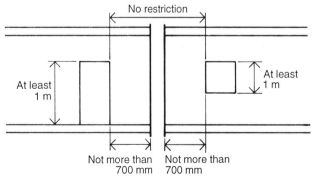

Figure 56

Northern Ireland, Technical Booklet G

The guidance under this heading complies with Technical Booklet G, June 1990, the Building Regulations (Northern Ireland) 1990. It is anticipated that future harmonisation with the Regulations for England and Wales will eliminate minor differences.

Solid masonry separating walls

Wall specifications in Technical Booklet G are as shown on page 36, with the following exceptions:

The mass requirement for construction D is 415 kg/m² excluding the plasterboard.

Technical Booklet G does not state that laminates of plasterboard and mineral wool may be used wherever plasterboard or plasterboard finish is specified.

Control of flanking transmission

Specifications in Technical Booklet G are all as shown on page 37, with the following exceptions:

At the junction between separating wall and ground or intermediate floors, it is not permitted to set joists into the separating wall. Joist hangers must be used.

For external walls, Technical Booklet G does not require the use of a cavity closer on the line of the separating wall. See Figure 57.

There must be a distance of 650 mm between openings which are on opposite sides of the separating wall. If the mass of the inner leaf is less than 120 kg/m², the requirement for openings within 700 mm applies as in England and Wales. See Figure 58.

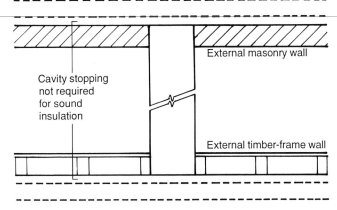

Figure 57

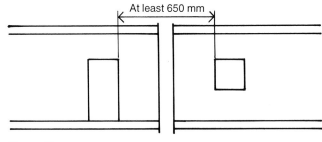

Figure 58

Scotland, Part H

The guidance under this heading complies with Part H of the Building Standards (Scotland) Regulations 1991. At the time of writing, Part H is under review. It is anticipated that harmonisation with the Regulations for England and Wales will eliminate minor differences.

Solid masonry separating walls

Wall specifications in Part H are as shown on page 36, with the following exceptions:

The mass requirement for construction D is 415 kg/m² excluding the plasterboard.

Part H does not state that laminates of plasterboard and mineral wool may be used wherever plasterboard or plasterboard finish is specified.

Control of flanking transmission

Specifications in Part H are as shown on page 37, with the following exceptions:

At the junction between separating wall and ground or intermediate floors, it is not permitted to set joists into the separating wall. Joist hangers must be used.

For external walls, Part H does not require the use of a cavity closer on the line of the separating wall. See Figure 59.

If the mass of the inner leaf of the external wall is less than 120 kg/m², the requirement for openings within 700 mm applies as in England and Wales.

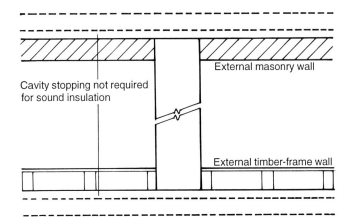

Figure 59

Other solid masonry separating wall constructions

Alternative constructions are permitted under the Regulations for England and Wales, Northern Ireland and Scotland, but only if it can be demonstrated that a given numerical standard will be, or has been, achieved. The designer should make sure that any alternative construction will meet the numerical field or laboratory test requirements for the relevant country.

Detailed guidance is given on pages 16 to 19. In England and Wales and in Northern Ireland, test evidence must be obtained before construction. A trade organisation or manufacturer may be able to provide suitable information. The designer should ensure that the test results were obtained in circumstances as close as possible to those in the new application.

In Scotland, post-construction testing is acceptable.

Walls of lower performance

Reducing the MASS of any solid masonry wall will result in reduced AIRBORNE SOUND INSULATION. The CRITICAL FREQUENCY dip also has an effect. The principles have been explained on page 14. Table 15 gives some practical examples.

Table 15

Construction	Approximate R_w (dB)*
Brickwork walls (field tests, R'_w)	
102.5 mm, plastered both sides (268 kg/m²)	47
215 mm, plastered both sides (488 kg/m²)	55
330 mm, plastered both sides (708 kg/m²) (with 450 mm brickwork flanking walls).	57
Autoclaved aerated concrete blockwork walls (laboratory tests, R_w)	
115 mm, fair face (73 kg/m²)	39
115 mm, dry-lined (95 kg/m²)	47
115 mm, plastered (93 kg/m²)	43
150 mm, fair-face (95 kg/m²)	42
150 mm, dry-lined (117 kg/m²)	48
150 mm, plastered (115 kg/m²)	46

* $R_w = D_{nT,w}$ when $S = 0.32 \times V$

where S = surface area of partition (m²)

and V = receiving room volume (m²)

See page 13 for full formulae

Walls of higher performance

Increasing the MASS of any solid masonry wall will result in increased AIRBORNE SOUND INSULATION of the wall element. FLANKING TRANSMISSION should be controlled by increasing masonry flanking walls in like proportion, or by carrying the separating wall through the flanking wall. The principles have been explained on page 14 (see Table 15).

Checklists
Design
● Set the MASS high enough.

● As the mass of the wall is reduced, the CRITICAL FREQUENCY increases to a point where its effects are more serious (see page 14).

● Specify that bricks be laid frog-up and all joints be fully filled with mortar.

● Specify a plaster finish to seal acoustically-porous blocks.

● Dry linings and other panel finishes can cause reduced low-frequency performance as a result of the MASS-AIR RESONANCE, particularly on dense concrete walls (see page 15).

● Observe the guidance given on pages 37 and 38 regarding window spacing in external flanking walls.

● Avoid areas of reduced MASS in the separating wall, for example electrical sockets which should not be installed back-to-back.

● Joists should preferably span parallel to the separating wall.

Site inspection
● Ensure that the bricks or blocks on site are of the correct density.*

● Ensure that the joints are fully filled with mortar and that bricks are laid frog-up.

● Check all junctions with surrounding constructions to ensure that there is an airtight seal.

● Look for areas of reduced mass and ensure that these are airtight and the reduction in mass is minimal.*

● Check the means of support for joists on the separating wall. Joist hangers are preferred.

● Inspect the separating wall in the roof space and where it passes through an intermediate floor.*

● Ensure that the plaster finish thickness is as specified.*

*   Check with job specification for requirements before inspecting.

Information sources
(See Appendix F)
AACPA
BCA
BDA
BRE

Cavity masonry walls

Factors affecting performance

The AIRBORNE SOUND INSULATION of cavity masonry walls is controlled mainly by MASS, ISOLATION, QUALITY OF CONSTRUCTION and FLANKING TRANSMISSION. Cavity masonry separating walls do not perform consistently better than solid walls of similar materials and mass for the following reasons:

● The individual leaves of the wall may be subject to a CRITICAL FREQUENCY dip in an important part of the frequency range (see page 14).

● The cavities are usually narrow and bridged by ties and at the edges. Consequently, the two leaves are not well isolated.

Construction A   (Figure 60(a))
Cavity brickwork plastered both sides

Mass   415 kg/m² or more (including plaster)

Isolation 50 mm cavity. Butterfly-pattern ties spaced as required for structural purposes*

Quality   Bricks laid frog-up. Joints fully filled and sealed with mortar. Junctions with surrounding constructions sealed

Example   102 mm leaves, brick density 1970 kg/m³, 75 mm coursing, 13 mm lightweight plaster

Construction B   (Figure 60(b))
Cavity concrete blockwork plastered both sides

Mass   415 kg/m² or more (including plaster)

Isolation 50 mm cavity. Butterfly-pattern ties spaced as required for structural purposes*

Quality   Joints fully filled with mortar. Junctions with surrounding constructions sealed

Example   100 mm leaves, block density 1990 kg/m³, 225 mm coursing, 13 mm lightweight plaster

Construction C   (Figure 60(c))
Cavity lightweight aggregate concrete blockwork

Mass   300 kg/m² or more, including 13 mm plaster or 12.5 mm plasterboard on both room faces. Density not more than 1600 kg/m³

Isolation 75 mm cavity. Butterfly-pattern ties spaced as required for structural purposes*

Quality   Joints fully filled with mortar. Junctions with surrounding constructions sealed, including those behind linings. Plasterboard joints sealed

Example   100 mm leaves, block density 1371 kg/m³, 225 mm coursing, 13 mm lightweight plaster

England and Wales, Approved Document E

The constructions which appear under this heading comply with Approved Document E, 1992 Edition, The Building Regulations 1991. It is not necessary to demonstrate that these constructions will meet a given numerical performance standard. The mean performance figures are given only to assist the designer.

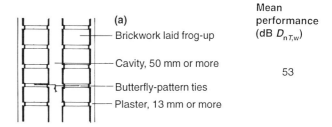

(a)
Brickwork laid frog-up
Cavity, 50 mm or more
Butterfly-pattern ties
Plaster, 13 mm or more

Mean performance (dB $D_{nT,w}$)

53

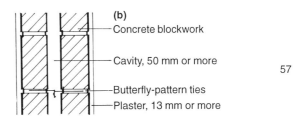

(b)
Concrete blockwork
Cavity, 50 mm or more
Butterfly-pattern ties
Plaster, 13 mm or more

57

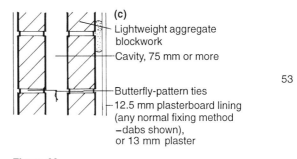

(c)
Lightweight aggregate blockwork
Cavity, 75 mm or more
Butterfly-pattern ties
12.5 mm plasterboard lining (any normal fixing method –dabs shown), or 13 mm plaster

53

Figure 60

Additional constructions for use only where a step and/or stagger of at least 300 mm is provided

Construction D   (Figure 61(a))
Cavity concrete blockwork with plasterboard linings

Mass   415 kg/m² or more for masonry alone, plus 12.5 mm plasterboard on both sides

Isolation 50 mm cavity. Butterfly-pattern ties spaced as required for structural purposes*

Quality   Joints fully filled with mortar. Junctions with surrounding constructions sealed, including those behind linings. Plasterboard joints sealed

Example   100 mm leaves, block density 1990 kg/m³, 225 mm coursing

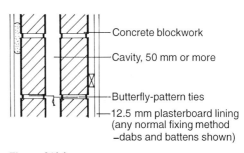

Concrete blockwork
Cavity, 50 mm or more
Butterfly-pattern ties
12.5 mm plasterboard lining (any normal fixing method –dabs and battens shown)

Figure 61(a)

Construction E (Figure 61(b))
Cavity lightweight aggregate concrete blockwork

Mass   250 kg/m² or more, including 13 mm plaster or 12.5 mm plasterboard on both room faces. Density not more than 1600 kg/m³

Isolation   75 mm cavity. Butterfly-pattern ties spaced as required for structural purposes*

Quality   Joints fully filled with mortar. Junctions with surrounding constructions sealed, including those behind linings. Plasterboard joints sealed

Example   100 mm leaves, block density 1105 kg/m³, 225 mm coursing, 13 mm lightweight plaster

* British Standard BS 5628 limits this tie type and spacing to cavities of 50 mm to 75 mm, with a minimum masonry leaf thickness of 90 mm. (BS 5628: Code of practice for use of masonry. Part 3:1985. Materials and components, design and workmanship.)

Control of flanking transmission
In section
(Figure 62)

The joint between the wall and the roof should be filled. A filling used for fire stopping, such mortar bedding or mineral fibre, should be used.

In the roof space, the wall mass may be reduced to not less than 150 kg/m², but only if the ceiling mass is equivalent to at least 12.5 mm plasterboard and the joints are well sealed. The cavity should still be maintained. If lightweight aggregate blocks of density less than 1200 kg/m³ are used, at least one side should be sealed using cement paint or plaster skim.

At the junction between separating wall and ground or intermediate floors, floor joists may be supported on hangers or built in. They may be built in only if good workmanship can be assured, and care should be taken to ensure that there are no airpaths through the wall. A suspended concrete intermediate or ground floor should be carried through to the cavity face of each leaf. A concrete slab on the ground may be continuous.

In plan
(Figure 63)

The outer leaf of a cavity wall may be of any construction. Where an external cavity wall has an inner leaf of masonry its mass should be at least 120 kg/m², except in the case of construction B on page 40 as shown in Figure 60(b), where there is no minimum required mass. Also:

● the masonry of the walls should be bonded together, or

● the masonry of the external wall should abut the separating wall and be tied to it with ties at no more than 300-mm centres vertically.

Where a cavity wall has an inner leaf of timber construction it should:

● abut the separating wall,

● be tied to it with ties at no more than 300-mm centres vertically, and

● have the joints sealed with tape or caulking.

The cavity in the separating wall should not be stopped by any material which connects the leaves rigidly together. Mineral wool is acceptable.

Partitions
There are no restrictions on partition walls meeting a cavity masonry separating wall.

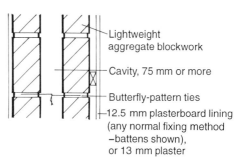

Figure 61(b)

Notes

Where plaster or plasterboard finish is specified, wall lining laminates of plasterboard and mineral fibre may be used instead.

If external walls are to be filled with an insulating material other than unbonded particles or fibres, the insulating material should be prevented from entering separating wall cavity by a flexible closer.

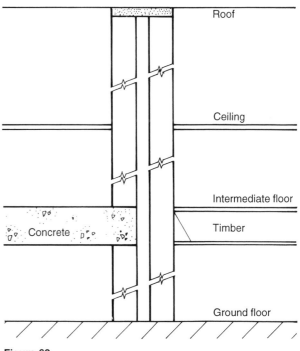

Figure 62

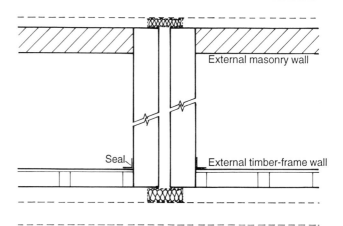

Figure 63

Northern Ireland, Technical Booklet G

The guidance under this heading complies with Technical Booklet G, June 1990, the Building Regulations (Northern Ireland) 1990. It is anticipated that future harmonisation with the Regulations for England and Wales will eliminate minor differences.

Cavity masonry separating walls

Wall specifications in Technical Booklet G are as shown on pages 40 and 41 with the following exceptions:

Construction C (Figure 60(c)) does not appear.

For construction E (Figure 61(b)), block densities up to 1500 kg/m³ only are permitted, and Technical Booklet G requires that the face of the blockwork be sealed with cement paint or plaster through the full width and depth of any intermediate floor.

For cavities up to 75 mm, Technical Booklet G requires butterfly-pattern wall ties to be spaced at a maximum of 900 mm apart horizontally and 450 mm apart vertically.

Technical Booklet G does not state that laminates of plasterboard and mineral wool may be used wherever plaster or plasterboard finish is specified.

Control of flanking transmission

Specifications in Technical Booklet G are all as shown on page 41 with the following exceptions:

At the junction between separating wall and ground or intermediate floors, it is not permitted to set joists into the separating wall. Joist hangers must be used.

There must be a distance of 650 mm between openings which are on opposite sides of the separating wall (see Figure 64).

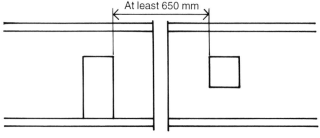

Figure 64

Scotland, Part H

The guidance under this heading complies with Part H of the Building Standards (Scotland) Regulations 1991. At the time of writing, Part H is under review. It is anticipated that harmonisation with the Regulations for England and Wales will eliminate minor differences.

Cavity masonry separating walls

Wall specifications in Part H are as shown on pages 40 and 41 with the following exceptions:

Construction C (Figure 60(c)) does not appear.

For construction E (Figure 61(b)), block densities up to 1500 kg/m³ only are permitted, and Part H requires that the face of the blockwork be sealed with cement paint or plaster through the full width and depth of any intermediate floor.

For cavities up to 75 mm, Part H requires butterfly-pattern wall ties to be spaced at a maximum of 900 mm apart horizontally and 450 mm apart vertically.

Part H does not state that laminates of plasterboard and mineral wool may be used wherever plaster or plasterboard finish is specified.

Control of flanking transmission

Specifications in Part H are as shown on page 41 with the following exceptions:

At the junction between separating wall and ground or intermediate floors, it is not permitted to set joists into the separating wall. Joist hangers must be used.

The cavity in the separating wall must be sealed only in accordance with Part D (Fire) which states the methods for fire stopping.

Other cavity masonry separating wall constructions

Alternative constructions are permitted under the Regulations for England and Wales, Northern Ireland and Scotland, but only if it can be demonstrated that a given numerical standard will be, or has been, achieved. The designer should make sure that any alternative construction will meet the numerical field or laboratory test requirements for the relevant country.

Detailed guidance is given on pages 16 to 19.

In England and Wales and in Northern Ireland, test evidence must be obtained before construction. A trade organisation or manufacturer may be able to provide suitable information. The designer should ensure that the test results were obtained in circumstances as close as possible to those in the new application.

In Scotland, post-construction testing is acceptable.

Other wall construction
Cavity autoclaved aerated concrete block, plastered
(Figure 65)

Mass   150 kg/m² or more (including plaster)

Isolation   Cavity 75 mm or more. Absorbent block face in cavity

Quality   Joints to be fully filled with mortar. Seal all junctions with surrounding construction

(This construction does not appear in any of the regulation documents for England and Wales, Northern Ireland or Scotland. Obtain suitable field test results before proceeding (see page 16), and consult with the block manufacturer on suitable edge details.)

Walls of lower performance
Cavity masonry walls are unlikely to be specified in situations where a lower degree of sound insulation is required.

Walls of higher performance
The airborne sound insulation can be improved by increasing the MASS or improving the ISOLATION between the leaves. The principles are explained on page 14.

Mass   Increasing the MASS of each leaf will improve performance and should push the CRITICAL FREQUENCY down, making its effects less serious

Isolation   Removal of the ties may improve the sound insulation of masonry walls by approximately 2 dB

Flanking transmission For best results, neither the external wall nor any floor construction should be carried across the cavity

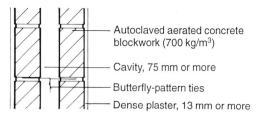

Figure 65

Autoclaved aerated concrete blockwork (700 kg/m³)

Cavity, 75 mm or more

Butterfly-pattern ties

Dense plaster, 13 mm or more

Checklists
Design
● Set the MASS high enough.

● Specify an appropriate cavity width.

● Specify only butterfly-pattern wire ties and do not space them too closely together.

● Specify that bricks be laid frog-up and that all joints be fully filled with mortar.

● Avoid areas of reduced MASS, for example electrical sockets which should not be installed back-to-back.

● Dry linings and other panel finishes can reduce low-frequency performance as a result of the MASS-AIR RESONANCE, particularly on dense concrete walls.

● Joists should preferably span parallel to the separating wall.

Site inspection
● Ensure that the bricks or blocks on site are of the correct density.*

● Ensure that only butterfly ties are being used and that they are correctly spaced.*

● Ensure that the specified cavity width is maintained and that the cavities are not bridged by any unspecified material.*

● Ensure that the joints are fully filled with mortar and that bricks are laid frog-up.

● Check all junctions with surrounding constructions to ensure that there is an airtight seal.

● Check that areas of reduced mass are airtight and that the reduction in mass is minimal.*

● Inspect the separating wall in the roof space and where it passes through an intermediate floor.*

● Ensure that the plaster finish thickness is at least as specified.*

* Check with job specification for requirements before inspecting.

Information sources
(See Appendix F)
AACPA
BCA
BDA
BRE

Masonry walls between isolated panels

Factors affecting performance

The AIRBORNE SOUND INSULATION of masonry between lightweight panels is controlled mainly by the MASS of the masonry and panels, the ISOLATION between them, QUALITY OF CONSTRUCTION and FLANKING TRANSMISSION.

Porous-faced blocks provide absorption in the cavity and so increase ISOLATION (see page 14). Consequently, their MASS can be reduced from that of 'harder' materials.

Masonry core constructions

Core type A (Figure 66(a))
Brickwork

Mass   300 kg/m² or more (excluding panels)

Quality   Bricks laid frog-up. Joints fully filled and sealed with mortar. Junctions with surrounding constructions sealed

Example   215 mm core, brick density 1290 kg/m³, 75 mm coursing

Core type B (Figure 66(b))
Concrete blockwork

Mass   300 kg/m² or more (excluding panels)

Quality   Joints fully filled and sealed with mortar. Junctions with surrounding constructions sealed

Example   140 mm core, block density 2200 kg/m³, 110 mm coursing

Core type C (Figure 66(c))
Lightweight concrete blockwork

Mass   160 kg/m² or more (excluding panels). Density not more than 1600 kg/m³

Quality   Joints fully filled and sealed with mortar. Junctions with surrounding constructions sealed

Example   200 mm core, block density 730 kg/m³, 225 mm coursing

Core type D (Figure 66(d))
Cavity brickwork or blockwork

Mass   No restriction, but each leaf at least 100 mm thick.

Isolation   50 mm cavity. Butterfly-pattern ties spaced as required for structural purposes

Quality   Bricks laid frog-up. Joints fully filled and sealed with mortar. Junctions with surrounding constructions sealed

Panel constructions

Panel construction E (Figure 66(e))
Plasterboard/cellular core

Mass   18 kg/m² or more (including any plaster finish)

Quality   Joints taped between panels and at edges

Isolation   Panel spaced 25 mm or more from the masonry core, and fixed to floor and ceiling only

Panel construction F (Figure 66(f))
Plasterboard/cellular core

Mass   Thickness of each sheet 12.5 mm if a supporting framework is used, or total thickness of at least 30 mm if no framework is used

Quality   Joints between sheets staggered

Isolation   Panel spaced 25 mm or more from the masonry core, and fixed to floor and ceiling only

England and Wales, Approved Document E

The constructions which appear under this heading comply with Approved Document E, 1992 Edition, The Building Regulations 1991. It is not necessary to demonstrate that these constructions will meet a given numerical performance standard. The mean performance figures are given only to assist the designer.

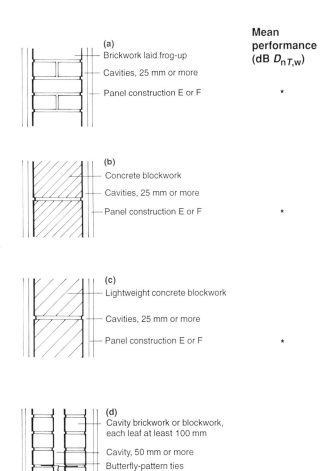

Mean performance (dB $D_{nT,w}$)

(a)
- Brickwork laid frog-up
- Cavities, 25 mm or more
- Panel construction E or F

*

(b)
- Concrete blockwork
- Cavities, 25 mm or more
- Panel construction E or F

*

(c)
- Lightweight concrete blockwork
- Cavities, 25 mm or more
- Panel construction E or F

*

(d)
- Cavity brickwork or blockwork, each leaf at least 100 mm
- Cavity, 50 mm or more
- Butterfly-pattern ties
- Cavities, 25 mm or more
- Panel construction E or F

*

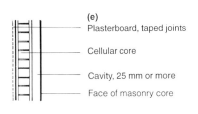

(e)
- Plasterboard, taped joints
- Cellular core
- Cavity, 25 mm or more
- Face of masonry core

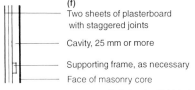

(f)
- Two sheets of plasterboard with staggered joints
- Cavity, 25 mm or more
- Supporting frame, as necessary
- Face of masonry core

* Extensive field test results are not available for this construction. It can be expected to meet or exceed the numerical values given on page 18, however.

Figure 66

Control of flanking transmission
In section
(Figure 67)

The joint between the wall and the roof should be filled. A filling used for fire stopping, such mortar bedding or mineral fibre, should be used.

In the roof space, the linings may be omitted and the wall mass may be reduced to not less than 150 kg/m², but only if the ceiling mass is equivalent to at least 12.5 mm plasterboard and the joints are well sealed. In the case of core type D (Figure 66(d)), the cavity should still be maintained. If lightweight aggregate blocks of density less than 1200 kg/m³ are used, at least one side should be sealed using cement paint or plaster skim. The junction between the freestanding panels and the ceiling should be sealed with tape or caulking.

At the junction between separating wall and ground or intermediate floors, hangers must be used to support any joists which span to the wall. Spaces between such joists should be filled with full depth timber blocking.

With a concrete intermediate floor, the floor base may only be carried through if it has a mass of at least 365 kg/m². If core type D is used, the cavity should not be bridged. The junction between the freestanding panels and the ceiling should be sealed with tape or caulking.

A concrete slab laid on the ground may be continuous. Concrete floors with a mass of less than 365 kg/m² should not bear on the core.

In plan
(Figure 68)

The outer leaf of a cavity wall may be of any construction.

With core type C (Figure 66(c)), the inner leaf of a cavity external wall should have an internal finish of isolated panels as shown for the separating walls. A layer of thermal insulation may be introduced to the cavity behind an isolated panel, provided that the cavity is not less than 25 mm wide.

The inner leaf of a cavity wall may be of any construction if it is lined with isolated panels.

Where the separating wall has core type A, B or D (Figure 66(a),(b) and (d)), plaster or dry lining with joints sealed with tape or caulking may be used. A layer of thermal insulation may be introduced to the cavity behind a dry lining, provided that the cavity is not less than 10 mm wide.

If the inner leaf of a cavity wall is plastered or dry-lined, it should have a total mass of 120 kg/m² and be butt-jointed to the core of the separating wall and tied to it with ties at no more than 300-mm centres vertically.

Partitions
Masonry partitions should not abut a separating wall of this type.

Loadbearing partitions should be fixed to the masonry core through a continuous padding of mineral-fibre quilt. The joint between the partition and the panels should be sealed using, for example, tape or caulking.

Non-loadbearing partitions should be tightly butted against lightweight panel. The joint should be sealed using, for example, mastic or jute scrim.

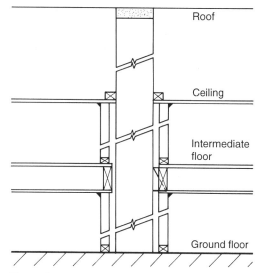

Figure 67

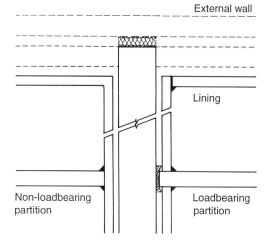

Figure 68

Northern Ireland, Technical Booklet G

The guidance under this heading complies with Technical Booklet G, June 1990, the Building Regulations (Northern Ireland) 1990. It is anticipated that future harmonisation with the Regulations for England and Wales will eliminate minor differences.

Masonry between isolated panels

Wall specifications in Technical Booklet G are as shown on page 44 with the following exceptions:

Core type C (Figure 66(c)) is valid only for autoclaved aerated concrete blockwork (see Figure 69). Lightweight aggregate concrete blockwork, up to a density of 1500 kg/m³, must achieve a mass of 200 kg/m².

Example 140 mm core, block density 1405 kg/m³, 225 mm coursing

Core type D (Figure 66(d)) does not appear in Technical Booklet G.

Technical Booklet G requires that any framing which supports the isolated panels should be kept 5 mm clear of the masonry core.

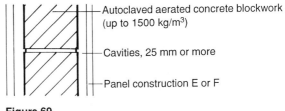

Figure 69

Autoclaved aerated concrete blockwork (up to 1500 kg/m³)

Cavities, 25 mm or more

Panel construction E or F

Control of flanking transmission

Specifications in Technical Booklet G are all as shown on page 45 with the following exceptions:

Technical Booklet G requires that the ground floor shall be a solid slab, laid on the ground.

There must be a distance of 650 mm between openings which are on opposite sides of the separating wall (Figure 70).

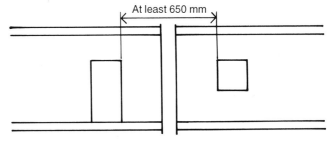

At least 650 mm

Figure 70

Scotland, Part H

The guidance under this heading complies with Part H of the Building Standards (Scotland) Regulations 1991. At the time of writing, Part H is under review. It is anticipated that harmonisation with the Regulations for England and Wales will eliminate minor differences.

Masonry between isolated panels

Wall specifications in Part H are as shown on page 44 with the following exceptions:

Core type C (Figure 66(c)) is valid only for autoclaved aerated concrete blockwork (see Figure 71). Lightweight aggregate concrete blockwork, up to a density of 1500 kg/m³, must achieve a mass of 200 kg/m².

Example 140 mm core, block density 1405 kg/m³, 225 mm coursing

Core type D (Figure 66(d)) does not appear in Part H.

Part H requires that any framing which supports the isolated panels should be kept 5 mm clear of the masonry core.

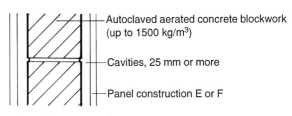

Figure 71

Autoclaved aerated concrete blockwork (up to 1500 kg/m³)

Cavities, 25 mm or more

Panel construction E or F

Control of flanking transmission

Specifications in Part H are all as shown on page 45 with the following exceptions:

Part H requires that the ground floor shall be a solid slab, laid on the ground.

Other masonry walls between isolated panels separating wall constructions

Alternative constructions are permitted under the Regulations for England and Wales, Northern Ireland and Scotland but only if it can be demonstrated that a given numerical standard will be, or has been, achieved. The designer should make sure that any alternative construction will meet the numerical field or laboratory test requirements for the relevant country.

Detailed guidance is given on pages 16 to 19 .

In England and Wales and in Northern Ireland, test evidence must be obtained before construction. A trade organisation or manufacturer may be able to provide suitable information. The designer should ensure that the test results were obtained in circumstances as close as possible to those in the new application.

In Scotland, post-construction testing is acceptable.

Walls of lower performance

Masonry walls with lightweight panels are unlikely to be specified in situations where a lower degree of sound insulation is required, as there are a number of other simpler methods available.

Walls of higher performance

The airborne sound insulation can be improved by increasing the MASS of the elements or by improving the ISOLATION between them. Where acoustically 'harder' masonry has been used, a fibrous absorber in the cavity will give a small increase in sound insulation by improving ISOLATION.

Flanking transmission can be reduced by installing the isolated panels next to the inside of any external or other flanking wall. Where possible, floor slabs should not be carried through the separating wall. Further reductions in flanking transmission via the floor slab could be made by:

● installing a floating floor on top of the slab (see pages 56 and 57), and/or

● installing a suspended or independent plasterboard ceiling below the slab (for example, see page 68).

Checklists
Design

● Specify the elements to achieve adequate MASS.

● Specify an appropriate cavity width.

● Do not permit anything to bridge the cavities.

● Specify that all joints be fully filled with mortar, and brickwork laid frog-up.

● Specify that all joints at the perimeter of the lightweight panel be well sealed, for example with tape, caulking or coving.

● Joists should span parallel to the separating wall. If this is not possible, use joist hangers.

Site inspection

● Ensure that the bricks or blocks on site are of the correct density.*

● Ensure that all joints are fully filled with mortar and that bricks are laid frog-up.

● Check all junctions with surrounding constructions to ensure that there is an airtight seal.

● *Most important:* ensure that the lightweight panels do not make physical contact with the masonry core and that the specified cavity width is maintained.*

● Inspect the separating wall in the roofspace and where it passes through an intermediate floor.

* Check with job specification for requirements before inspecting.

Information sources

(See Appendix F)
AACPA
BCA
BDA
BG
BRE

Timber-frame walls with absorbent curtain

Factors affecting performance

The AIRBORNE SOUND INSULATION of a timber-frame wall with absorbent curtain depends mainly on the ISOLATION between two lightweight claddings, their MASS, QUALITY OF CONSTRUCTION and FLANKING TRANSMISSION.

England and Wales, Approved Document E

The constructions which appear under this heading comply with Approved Document E, 1992 Edition, The Building Regulations 1991. It is not necessary to demonstrate that these constructions will meet a given numerical performance standard. The mean performance figures are given only to assist the designer.

Timber frame
In plan
(Figure 72)

Mass   At least two sheets of plasterboard, with staggered joints. Thickness 30 mm or more to both sides

Isolation   Cavity between claddings 200 mm or more. Absorbent mineral fibre (which may be wire-reinforced), density 10 kg/m^3 or more, thickness 25 mm if suspended in the cavity between the frames, 50 mm if fixed to one frame, or 25 mm per quilt if one fixed to both frames. Plywood sheathing may be used in the cavity as necessary for structural reasons

Quality   Plasterboard joints staggered and junctions with surrounding constructions sealed. If the frames must be connected together, 14–15 gauge (40 mm × 3 mm) metal straps should be fixed at or just below ceiling level 1.2 m apart. Power points may be set in the linings provided there is a similar thickness of cladding behind the socket box. Power points should not be placed back-to-back across the wall. Where fire stops are needed in the cavity between frames they should either be flexible or be fixed to one frame only.

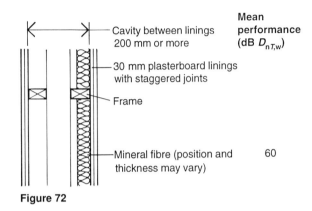

Cavity between linings 200 mm or more

30 mm plasterboard linings with staggered joints

Frame

Mineral fibre (position and thickness may vary)

Mean performance (dB $D_{nT,w}$)

60

Figure 72

Timber frames, masonry core
In plan
(Figure 73)

The above specification may be used with a masonry core, of any mass, in the cavity. Framing preferably should be clear of the core.

A masonry core does not normally improve sound resistance, but may be useful for support and in stepped or staggered situations. There are no restrictions on type but the core should be connected to only one frame.

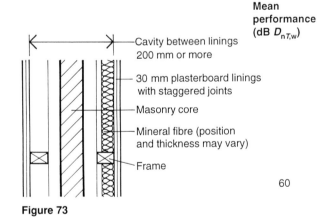

Mean performance (dB $D_{nT,w}$)

Cavity between linings 200 mm or more

30 mm plasterboard linings with staggered joints

Masonry core

Mineral fibre (position and thickness may vary)

Frame

60

Figure 73

Control of flanking transmission

In section
(Figure 74)

The joint between wall and roof should be fire-stopped (see Approved Document B, Fire Safety).

Between the ceiling level and the underside of the roof either:

● both frames should be carried through and at least 25 mm plasterboard fixed to each, or

● the cavity should be closed at ceiling level without connecting the two frames rigidly together, and one frame should be carried through with at least 25 mm plasterboard on both sides. (A suggested detail is given below.)

In each case, the space between the frame and the roof finish should be sealed. At an intermediate floor, the airpaths to the wall cavity should be blocked either by carrying the cladding through the floor or by using a solid timber edge to the floor. Where the joists are at right angles to the wall, spaces between joists should be sealed using full-depth timber blocking. The ground floor may be a ground bearing concrete slab or a suspended concrete slab. If suspended, the mass should be at least 365 kg/m^2.

In plan
(Figure 75)

Outer leaf/cladding: there are no restrictions.

If the external wall is a cavity wall, the cavity should be sealed between the ends of the separating wall and the outer leaf, to prevent soundpaths.

The internal finish should be 12.5 mm plasterboard or other equally heavy material. (Resilient layers for thermal insulation may be incorporated if desired.)

Partitions

There are no restrictions on internal partitions meeting a separating wall of this type.

Suggested details
(based on guidance in Approved Document E)

Where it is not feasible to carry both frames through, a cavity closer should be installed at ceiling level, for example fire-resistant board fixed rigidly to one frame and isolated from the other frame using a fibrous quilt packing.

The airpaths between the rooms and the wall cavity should be blocked using timber noggings, as shown in Figure 76, or the plasterboard linings carried through.

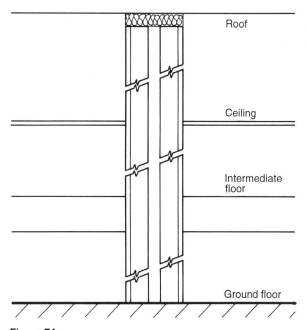

Figure 74

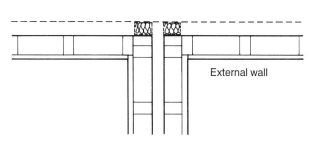

Figure 75

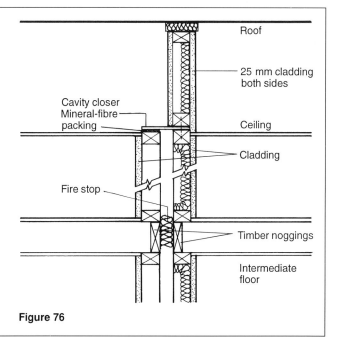

Figure 76

49

Northern Ireland, Technical Booklet G

The guidance under this heading complies with Technical Booklet G, June 1990, the Building Regulations (Northern Ireland) 1990. It is anticipated that future harmonisation with the Regulations for England and Wales will eliminate minor differences.

Timber frames with absorbent curtain

Wall specifications in Technical Booklet G are as shown on page 48 with the following exceptions:

Services are not to be contained in the wall, so avoiding the creation of airpaths through the lining.

The absorbent curtain should be unfaced mineral fibre (which may be wire-reinforced), density 12 to 36 kg/m³.

Frames are to be connected only if necessary for structural reasons, and then as few ties as possible should be used, not more than 14–16 gauge (40 mm × 3 mm) metal straps fixed at or just below ceiling level, 1.2 m apart.

For timber frames with masonry core (Figure 73), Technical Booklet G requires that the framing be kept 5 mm clear of the masonry core.

Control of flanking transmission

Specifications in Technical Booklet G are all as shown on page 49 with the following exceptions:

Technical Booklet G requires that the complete separating wall construction should be carried through to the underside of the roof covering, irrespective of the presence of any type of ceiling.

Technical Booklet G requires that the ground floor shall be a solid slab, laid on the ground.

There must be a distance of 650 mm between openings which are on opposite sides of the separating wall (Figure 77).

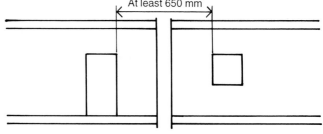

Figure 77

Scotland, Part H

The guidance under this heading complies with Part H of the Building Standards (Scotland) Regulations 1991. At the time of writing, Part H is under review. It is anticipated that harmonisation with the Regulations for England and Wales will eliminate minor differences.

Timber frames with absorbent curtain

Wall specifications in Part H are as shown on page 48 with the following exceptions:

This construction type is limited to use between dwellings up to three storeys high, subject to certain conditions (see Part D, Scotland).

Services are not to be contained in the wall, so avoiding the creation of airpaths through the lining.

The absorbent curtain should be unfaced mineral fibre (which may be wire-reinforced), density 12 to 36 kg/m³.

Frames are to be connected only if necessary for structural reasons, and then as few ties as possible should be used, not more than 14–16 gauge (40 mm × 3 mm) metal straps fixed at or just below ceiling level, 1.2 m apart.

For timber frames with masonry core (Figure 73), Part H requires that the framing be kept 5 mm clear of the masonry core.

Control of flanking transmission

Specifications in Part H are all as shown on page 49 with the following exceptions:

Part H requires that the complete separating wall construction should be carried through to the underside of the roof covering, irrespective of the presence of any type of ceiling.

Other timber-frame/absorbent curtain separating wall constructions

Alternative constructions are permitted under the Regulations for England and Wales, Northern Ireland and Scotland, but only if it can be demonstrated that a given numerical standard will be, or has been, achieved. The designer should make sure that any alternative construction will meet the numerical field or laboratory test requirements for the relevant country.

Detailed guidance is given on pages 16 to 19 .

In England and Wales and in Northern Ireland, test evidence must be obtained before construction. A trade organisation or manufacturer may be able to provide suitable information. The designer should ensure that the test results were obtained in circumstances as close as possible to those in the new application.

In Scotland, post-construction testing is acceptable.

Walls of lower performance

Where a lower standard of sound insulation can be accepted, for example for internal partitions, the ISOLATION between the linings could be reduced. These can be fixed to both sides of a single row of timber studs or steel studs. The following factors will control the sound insulation in practice (see also page 14):

Isolation   Metal studs give better isolation than timber studs. The isolation of timber studs can be improved by fixing one of the linings using, for example, resilient metal channels. An absorbent quilt in the cavity improves acoustic isolation.

Mass   Increasing the mass of the linings gives increased sound insulation. However, beyond approximately 30 kg/m^2 plasterboard (38 mm), the CRITICAL FREQUENCY dip falls below 1000 Hz, where its effects may be more serious.

Table 16 gives practical examples with their measured performance in the laboratory.

Table 16

	Construction	R_w (dB)*
1	Single layer 12.5 mm plasterboard	28
2	Two layers 12.5 mm plasterboard bonded together	31
3	75 mm timber studs with one layer of 12.5 mm plasterboard each side	36
4	As for 3, but with 25 mm absorbent quilt in the cavity	40
5	100 mm timber studs with two layers of 12.5 mm plasterboard both sides	45
6	As for 5, but with 25 mm absorbent quilt in the cavity	46
7	100 mm timber studs with two layers of 12.5 mm plasterboard, one side fixed via resilient channels, 25 mm absorbent quilt in the cavity	50
8	70 mm metal channel with two layers of 12.5 mm plasterboard both sides	48
9	As for 8, but with 25 mm absorbent quilt in the cavity	53
10	100 mm timber studs with wood-lath and plaster both sides	38

* $R_w = D_{nT,w}$ when $S = 0.32 \times V$

     where: S = partition surface area (m^2)
                   V = room volume (m^2)

   See page 13 for full formulae

Walls of higher performance

The above principles apply. Where sound insulation of more than 60 dB $D_{nT,w}$ is required, specialist advice should be obtained.

Checklists
Design

● Do not specify anything which bridges the cavity other than necessary metal straps at the recommended spacing, and suitably designed cavity closers.

● Avoid services penetrations through the linings.

● If socket outlets must be located on a separating wall, specify that the lining enclose the boxes, and do not position them back-to-back.

● Specify that all perimeter joints be well sealed, for example with mastic, jute scrim or coving.

Site inspection

● *Most important:* ensure that the specified cavity width is provided and that no unspecified items bridge the cavity.*

● Ensure that there are no unspecified penetrations of the linings.

● Ensure that any socket outlets in the separating wall are enclosed by the cladding material and not located back-to-back.

● Ensure that the absorbent quilt is of the specified thickness and has been fixed in the specified position.*

● Ensure that the plasterboard linings are of the specified thickness, the joints staggered and the perimeter well-sealed in accordance with the specification.*

*   Check with job specification for requirements before inspecting.

Information sources
(See Appendix F)
BG
BRE
TRADA

Suspended concrete floor with a soft covering

Factors affecting performance

The AIRBORNE SOUND INSULATION of a suspended concrete floor with a soft covering is controlled mainly by the MASS of the concrete base, the QUALITY OF CONSTRUCTION and FLANKING TRANSMISSION.

The IMPACT SOUND INSULATION of a suspended concrete floor with a soft covering is controlled mainly by the soft covering which reduces impact sound at source, the MASS of the concrete base and FLANKING TRANSMISSION.

England and Wales, Approved Document E

The constructions which appear under this heading comply with Approved Document E, 1992 Edition, The Building Regulations 1991. It is not necessary to demonstrate that these constructions will meet the numerical performance standards associated with each of the Regulations. The mean performance figures are given only to assist the designer.

Floor bases

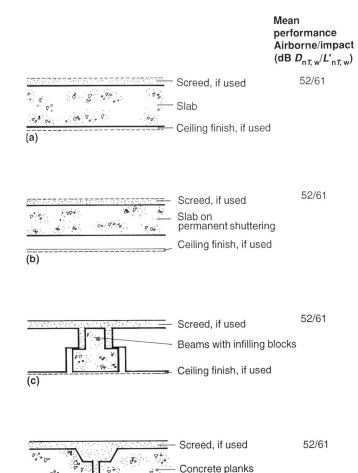

Figure 78

Mean performance Airborne/impact (dB $D_{nT,w}/L'_{nT,w}$)

Floor base A
Solid concrete slab (in-situ)
(Figure 78(a))

Mass — 365 kg/m² including any screed and/or ceiling finish

Quality — All joints between parts of the floor base filled to avoid airpaths

52/61

Floor base B
Solid concrete slab with permanent shuttering
(Figure 78(b))

Mass — 365 kg/m² including any screed and/or ceiling finish, and including shuttering if it is solid concrete or metal

Quality — All joints between parts of the floor base filled to avoid airpaths

52/61

Floor base C
Concrete beams with infilling blocks
(Figure 78 (c))

Mass — 365 kg/m² including any screed and/or ceiling finish

Quality — Floor surface level (a levelling screed may be necessary). All joints between parts of the floor base filled to avoid airpaths

52/61

Floor base D
Concrete planks (solid or hollow)
(Figure 78(d))

Mass — 365 kg/m² including any screed and/or ceiling finish

Quality — Floor surface level (a levelling screed may be necessary). All joints between parts of the floor base filled to avoid airpaths

52/61

Soft covering

Any resilient material, or material with a resilient base, with an overall uncompressed thickness of at least 4.5 mm

Suitable resilience will also be provided by a floor covering with a weighted impact sound improvement (ΔL_w) of not less than 17 dB, as calculated in Annex A to British Standard BS 5821: Part 2:1984 (see page 13).

Control of flanking transmission
External wall or cavity separating wall
The mass of the leaf adjoining the floor should be 120 kg/m²
(including any finish), unless it is an external wall having openings of
at least 20% of its area in each room, in which case there is no
minimum requirement.

(Openings in the external wall on either side of the separating floor
have the effect of restricting the flow of energy between the two
portions of the flanking wall. This reduces flanking transmission.)

The floor base (excluding any screed, even in floor bases C and D)
should pass through the leaf, whether spanning parallel to or at right
angles to the wall. The cavity should not be bridged. If the floor base
is type C or D, where the beams are parallel to the wall, the first
joint should be at least 300 mm from the cavity face of the wall leaf
(Figure 79).

Internal wall or solid separating wall
(Figure 80)
If the wall mass is less than 375 kg/m² including any plaster or
plasterboard, then the floor base excluding any screed should pass
through.

If the wall mass is more than 375 kg/m² including any plaster or
plasterboard, either the wall excluding any finishes or the floor base
excluding any screed may pass through. Where the wall does pass
through, the floor base should be tied to the wall and the joint
grouted.

Floor penetrations (excluding gas pipes)
(Figure 81)
Ducts or pipes penetrating a floor separating habitable rooms should
be in an enclosure, both above and below the floor.

Openings in the slab should be no larger than necessary.

The material of the enclosure should have a mass of 15 kg/m².

Either the enclosure should be lined or the duct or pipe should be
wrapped using 25 mm unfaced mineral fibre.

Penetrations through a separating floor by ducts and pipes should
have fire protection in accordance with Approved Document B: Fire
safety. The fire-stopping should be flexible and should prevent rigid
contact between pipe and floor.

Floor penetrations for gas pipes
In the Gas Safety Regulations 1972 (SI. 1972/1178) and the Gas
Safety (Installation and use) Regulations 1984 (SI. 1984/1358) there
are requirements for the ventilation of ducts at each floor where
they contain gas pipes. Gas pipes may be contained in a separate
ventilated duct or they can remain unducted.

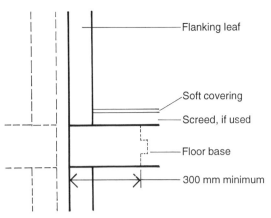

Figure 79

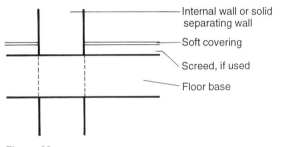

Figure 80

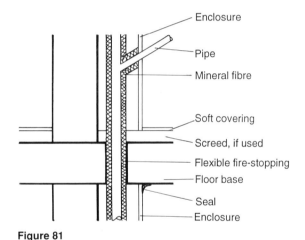

Figure 81

Northern Ireland, Technical Booklet G

The guidance under this heading complies with Technical Booklet G, June 1990, The Building Regulations (Northern Ireland) 1990. It is anticipated that future harmonisation with the Regulations for England and Wales will eliminate minor differences.

Suspended concrete floor with a soft covering

Floor constructions in Technical Booklet G are as shown on page 52 with the following exceptions:

With floor bases C and D (see page 52), a floor screed or structural topping is required.

Soft covering must be bonded to the floor base.

Control of flanking transmission

Specifications in Technical Booklet G are as shown on page 53 with the following exceptions:

External wall or cavity separating wall
The mass requirement of 120 kg/m^2 includes any plaster, but not any other finish. For floor bases C and D (see page 52), where the joints are parallel to the wall, Technical Booklet G does not stipulate a minimum distance for the first joint from the wall.

Internal wall or solid separating wall
The wall mass below which the floor base should pass through the wall is 355 kg/m^2 (including any plaster) not 375 kg/m^2.

Floor penetrations
The requirements are not restricted to habitable rooms. Where a flue pipe penetrates the floor and does not discharge into a flue within a chimney carried by the floor, a non-combustible enclosing duct with mineral-fibre absorbent lining shall be provided.

Scotland, Part H

The guidance under this heading complies with Part H of the Building Standards (Scotland) Regulations 1991. At the time of writing, Part H is under review. It is anticipated that harmonisation with the Regulations for England and Wales will eliminate minor differences.

Suspended concrete floor with a soft covering

Floor constructions in Part H are as shown on page 52 with the following exceptions:

With floor bases C and D, a floor screed or structural topping must be used.

Soft covering must be bonded to the floor base.

Control of flanking transmission

Specifications in Part H are as shown on page 53 with the following exceptions:

External wall or cavity separating wall
The mass requirement of 120 kg/m^2 includes any plaster, but not any other finish. For floor bases C and D (see page 52), where the joints are parallel to the wall, Part H does not stipulate a minimum distance for the first joint from the wall.

Internal wall or solid separating wall
The wall mass below which the floor base should pass through the wall is 355 kg/m^2 (including any plaster), not 375 kg/m^2.

Floor penetrations
The requirements are not restricted to habitable rooms. Where permitted by Part F, Scotland, a flue pipe may penetrate the floor. Unless the flue pipe discharges into a flue within a chimney carried by the floor, a non-combustible enclosing duct with mineral-fibre absorbent lining must be provided.

Other suspended concrete floor with soft covering separating floor constructions

Alternative constructions are permitted under the Regulations for England and Wales, Northern Ireland and Scotland, but only if it can be demonstrated that a given numerical standard will be, or has been, achieved. The designer should make sure that any alternative construction will meet the numerical field or laboratory test requirements for the relevant country.

Detailed guidance is given on pages 16 to 19 .

In England and Wales and in Northern Ireland, test evidence must be obtained before construction. A trade organisation or manufacturer may be able to provide suitable information. The designer should ensure that the test results were obtained in circumstances as close as possible to those in the new application.

In Scotland, post-construction testing is acceptable.

Floors of lower performance

Reducing the MASS of the concrete floor will result in reduced AIRBORNE and IMPACT SOUND INSULATION. The CRITICAL FREQUENCY dip also has an effect. The physical principles are explained on page 14.

The IMPACT SOUND LEVEL also relates to the covering material. Some manufacturers provide laboratory test results for the REDUCTION OF IMPACT SOUND-PRESSURE LEVEL. This can be used to compare materials. Some practical examples are given on page 12.

Floors of higher performance

Increasing the MASS of the concrete floor will result in increased AIRBORNE and IMPACT SOUND INSULATION. However, improvements will often be limited because of flanking transmission. Flanking transmission can be minimised by carrying the floor base through flanking walls, by increasing the mass of masonry flanking walls in like proportion, or by installing an isolated wall lining in front of flanking walls, such as those specified on page 44.

For further IMPACT improvements, a covering material with a good REDUCTION OF IMPACT SOUND LEVEL should be specified. Some practical examples are given on page 12.

Checklists
Design
● Set the MASS high enough by specifying the concrete mix and thickness, or by appropriate selection of precast elements.

● In thin slabs, the CRITICAL FREQUENCY will rise to a point where its effects are more serious (see page 14).

● Specify that all joints should be well formed and free from honeycombing.

● Keep services penetrations to a minimum, and encase pipes.

● Ceilings which enclose a small airgap can cause reduced low-frequency performance due to the MASS-AIR RESONANCE (see page 15).

● All junctions should be well filled and sealed; check the specification of all junction details, and specify a screed to ensure the floor has no airpaths (see page 51).

● Build the floor in on all sides.

Site inspection
● Check the thickness of the suspended concrete slab and the mix for in-situ concrete, in particular the quantity and type of aggregate used.*

● Check all junctions with surrounding constructions and ensure that all joints are well formed and free from honeycombing.*

● Ensure that services penetrations are no larger than necessary and that any pipes are properly wrapped.*

● Ensure that the floor base is airtight and that, where specified, a screed has been provided.*

● Ensure that the floor is built-in on all sides, if specified.*

* Check with specification before inspecting.

Information sources
(See Appendix F)
BCA
BRE

Suspended concrete floor with a floating layer

Factors affecting performance

The AIRBORNE SOUND INSULATION of a suspended concrete floor with floating layer is controlled mainly by the MASS of the base floor, the ISOLATION between the floating layer and the concrete base, QUALITY OF CONSTRUCTION and FLANKING TRANSMISSION.

The IMPACT SOUND INSULATION of a concrete floor with floating layer is controlled by the ISOLATION between the floating layer and the base floor, the MASS of the floating layer and base floor, QUALITY OF CONSTRUCTION and FLANKING TRANSMISSION.

Floor bases

Floor base A   (Figure 82(a))
Solid concrete slab (in-situ)

Mass 300 kg/m² including any bonded screed and/or ceiling finish

Quality All joints between parts of the floor base filled to avoid airpaths

Floor base B   (Figure 82(b))
Solid concrete slab with permanent shuttering

Mass 300 kg/m² including any bonded screed and/or ceiling finish, and including shuttering if it is solid concrete or metal

Quality All joints between parts of the floor base filled to avoid airpaths

Floor base C   (Figure 82(c))
Concrete beams with infilling blocks

Mass 300 kg/m² including any bonded screed and/or ceiling finish

Quality Floor base reasonably level (5 mm maximum step between units). A levelling screed may be required. Joints filled to avoid airpaths

Floor base D   (Figure 82(d))
Concrete planks (solid or hollow)

Mass 300 kg/m² including any bonded screed and/or ceiling finish

Quality Floor base reasonably level (5 mm maximum step between units). A levelling screed may be required. Joints filled to avoid airpaths

Floating layers

Timber raft
(Figure 82(e))

Mass Tongue-and-groove boarding at least 18 mm thick fixed to 45 mm × 45 mm battens.

Quality Raft laid loose on resilient layer. ISOLATION should not be bridged by nailing through the battens to the floor base. Resilient layer turned up at the edges

Screed
(Figure 82(f))

Mass 65 mm cement sand screed with 20 mm to 50 mm wire mesh (to protect the resilient layer while the screed is being laid)

Quality Screed must not enter or penetrate the resilient layer. The resilient layer may be protected by a paper facing.

England and Wales, Approved Document E

The constructions which appear under this heading comply with Approved Document E, 1992 Edition, The Building Regulations 1991. It is not necessary to demonstrate that these constructions will meet the numerical performance standards associated with each of the Regulations. The mean performance figures are given only to assist the designer.

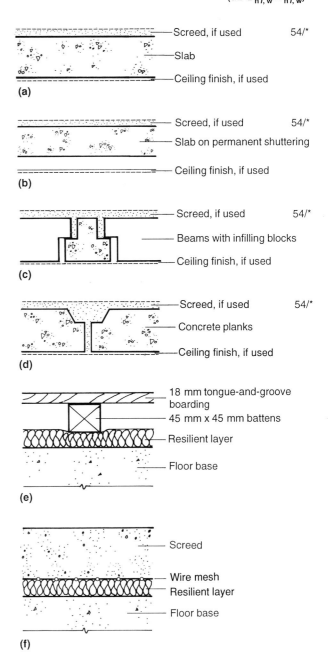

Mean performance Airborne/impact (dB $D_{nT,w}/L'_{nT,w}$)

(a) — Screed, if used — Slab — Ceiling finish, if used — 54/*

(b) — Screed, if used — Slab on permanent shuttering — Ceiling finish, if used — 54/*

(c) — Screed, if used — Beams with infilling blocks — Ceiling finish, if used — 54/*

(d) — Screed, if used — Concrete planks — Ceiling finish, if used — 54/*

(e) — 18 mm tongue-and-groove boarding — 45 mm x 45 mm battens — Resilient layer — Floor base

(f) — Screed — Wire mesh — Resilient layer — Floor base

* The impact sound insulation of these floors has proved variable in the field. If the advice on detailing is followed, the resulting impact insulation should be significantly better than the numerical values given on page 16.

Figure 82

56

Resilient layers
Mineral fibre

Isolation   Density 36 kg/m³. Thickness at least 25 mm (13 mm under a timber raft with battens which have an integral closed-cell resilient foam strip)

Quality   Fibre tightly butted, with turned-up edges at the floor perimeter. Under a screed, fibre should be paper-faced on the upper side to prevent screed entering the layer. Under a timber raft the fibre may be paper-faced on the underside

Pre-compressed expanded polystyrene board
(impact sound duty grade)
(Use under screeds only)

Isolation   Thickness 13 mm

Quality   Boards tightly butted; use on edge as a resilient strip at the perimeter of screed

Extruded polyethylene foam
(Use under screeds only)

Isolation   Density 30 to 45 kg/m³. Thickness 5 mm

Quality   Foam laid on a levelling screed to prevent puncture. Joints lapped and turned up at edges of the floating screed

Control of flanking transmission
External wall or cavity separating wall

The mass of the leaf adjoining the floor should be 120 kg/m² (including any finish), unless it is an external wall having openings of at least 20% of its area in each room, in which case there is no minimum requirement.

(Openings in the external wall on either side of the separating floor have the effect of restricting the flow of energy between the two portions of the flanking wall. This reduces flanking transmission.)

The floor base (excluding any screed) should pass through the leaf, whether spanning parallel to, or at right angles to, the wall. The cavity should not be bridged. If the floor base comprises concrete beams with infilling blocks or concrete planks, where the beams are parallel to the wall the first joint should be at least 300 mm from the cavity face of the wall leaf (Figure 83).

The resilient layer should be turned up at all edges to isolate the floating layer (Figure 84).

A nominal gap should be left between the skirting and the floating layer, or the resilient layer turned under the skirting. A seal is not necessary, but if used it should be flexible.

Internal wall or solid separating wall

If the wall mass is less than 375 kg/m² including any plaster or plasterboard, then the floor base excluding any screed should pass through.

If the wall mass is more than 375 kg/m² including any plaster or plasterboard, either the wall excluding any finishes or the floor base excluding any screed may pass through. Where the wall does pass through, the floor base should be tied to the wall and the joint grouted.

Floor penetrations (excluding gas pipes)
(Figure 85)
Ducts or pipes penetrating a floor separating habitable rooms should be in an enclosure, both above and below the floor.

Openings in the slab should be no larger than necessary.

The material of the enclosure should have a mass of 15 kg/m².

Either the enclosure should be lined, or the duct or pipe should be wrapped using 25 mm unfaced mineral fibre. A nominal gap should be left between the enclosure and the floating layer, and a seal should be provided, using acrylic caulking or neoprene.

Penetrations through a separating floor by ducts and pipes should have fire protection in accordance with Approved Document B: Fire safety. The fire-stopping should be flexible and should prevent rigid contact between pipe and floor.

Floor penetrations for gas pipes
In the Gas Safety Regulations 1972 (SI. 1972/1178) and the Gas Safety (Installation and use) Regulations 1984 (SI. 1984/1358) there are requirements for the ventilation of ducts at each floor where they contain gas pipes. Gas pipes may be contained in a separate ventilated duct or they can remain unducted.

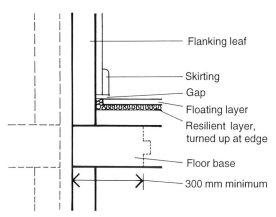

Figure 83

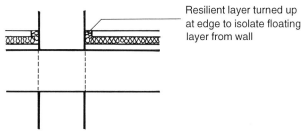

Figure 84

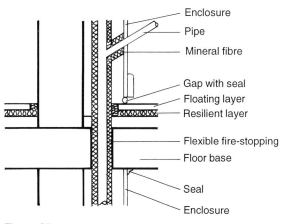

Figure 85

Northern Ireland, Technical Booklet G

The guidance under this heading complies with Technical Booklet G, June 1990, The Building Regulations (Northern Ireland) 1990. It is anticipated that future harmonisation with the Regulations for England and Wales will eliminate minor differences.

Suspended concrete floor/floating layer construction

Floor constructions in Technical Booklet G are as shown on pages 56 and 57 with the following exceptions:

Floor screeds and ceiling finishes are described as 'optional' for all base constructions.

With floor bases C and D no dimension is given for the maximum allowable step between units, but a levelling screed is recommended.

The resilient layer should be protected by a chicken-wire type of underlay (of unspecified dimensions) when a floating screed is being laid.

Control of flanking transmission

Specifications are as shown on page 57 with the following exceptions:

External wall or cavity separating wall
The mass requirement of 120 kg/m^2 includes any plaster, but not any other finish. For floor bases C and D (see page 56), where the joints are parallel to the wall, Technical Booklet G does not stipulate a minimum distance for the first joint from the wall. A 3 mm gap should be left between skirting and floating layer, or the resilient layer should be turned up under the skirting.

Internal wall or solid separating wall
The wall mass below which the floor base should pass through the wall is 355 kg/m^2 (including any plaster), not 375 kg/m^2.

Floor penetrations
The requirements are not restricted to habitable rooms. Where a flue pipe penetrates the floor and does not discharge into a flue within a chimney carried by the floor, a non-combustible enclosing duct with mineral-fibre absorbent lining shall be provided.

Scotland, Part H

The guidance under this heading complies with Part H of the Building Standards (Scotland) Regulations 1991. At the time of writing, Part H is under review. It is anticipated that harmonisation with the Regulations for England and Wales will eliminate minor differences.

Suspended concrete floor/floating layer construction

Floor constructions in Part H are as shown on pages 56 and 57 with the following exceptions:

Floor screeds and ceiling finishes are described as 'optional' for all base constructions.

With floor bases C and D, no dimension is given for the maximum allowable step between units, but a levelling screed is recommended.

The resilient layer should be protected by mesh underlay (of unspecified dimensions) when a floating screed is being laid.

Control of flanking transmission

Specifications are as shown on page 57 with the following exceptions:

External wall or cavity separating wall
The mass requirement of 120 kg/m^2 includes any plaster, but not any other finish. For floor bases C and D (see page 56), where the joints are parallel to the wall, Part H does not stipulate a minimum distance for the first joint from the wall. A 3 mm gap should be left between skirting and floating layer, or the resilient layer should be turned up under the skirting.

Internal wall or solid separating wall
The wall mass below which the floor base should pass through the wall is 355 kg/m^2 (including any plaster), not 375 kg/m^2.

Floor penetrations
The requirements are not restricted to habitable rooms. Where permitted by Part F of the Regulations, a flue pipe may penetrate the floor. Unless the flue pipe discharges into a flue within a chimney carried by the floor, a non-combustible enclosing duct with mineral-fibre absorbent must be provided.

Other suspended concrete floor with floating layer separating floor constructions

Alternative constructions are permitted under the Regulations for England and Wales, Northern Ireland and Scotland, but only if it can be demonstrated that a given numerical standard will be, or has been, achieved. The designer should make sure that any alternative construction will meet the numerical field or laboratory test requirements for the relevant country.

Detailed guidance is given on pages 16 to 19 .

In England and Wales and in Northern Ireland, test evidence must be obtained before construction. A trade organisation or manufacturer may be able to provide suitable information. The designer should ensure that the test results were obtained in circumstances as close as possible to those in the new application.

In Scotland, post-construction testing is acceptable.

Floors of lower performance

Reducing the MASS of the concrete floor will result in reduced AIRBORNE and IMPACT SOUND INSULATION. The CRITICAL FREQUENCY dip also has an effect. The physical principles are explained on page 14

Floors of higher performance

Increasing the MASS of the concrete base will result in increased AIRBORNE and IMPACT SOUND INSULATION. However, improvements will often be limited because of flanking transmission. Flanking transmission can be minimised by carrying the floor base through flanking walls, by increasing the mass of the masonry flanking walls in like proportion, or by installing an isolated wall lining in front of flanking walls, such as those specified on page 44.

For further IMPACT improvements, a soft floor covering with a good REDUCTION OF IMPACT SOUND LEVEL should be specified. Some practical examples are given on page 12.

Checklists
Design
- Set the MASS high enough by specifying the concrete mix and thickness, or by appropriate selection of precast elements.

- In thin slabs, the CRITICAL FREQUENCY will rise to a point where its effects are more serious (see page 14).

- Specify that all joints be well formed and free from honeycombing.

- Keep services penetrations to a minimum, and encase pipes.

- Specify no bridging between the floating layer and the concrete base, and do not run services under the floating floor.

- A floating timber raft can cause a dip in performance due to the MASS-AIR RESONANCE (see page 15). Ensure that the battens are deep enough.

- Ceilings which enclose a small airgap can cause reduced low-frequency performance due to the MASS-AIR RESONANCE (see page 15).

- Check the specification of all junction details.

- Check with the resilient layer manufacturer that the product is suitable for long-term wear under a timber raft construction.

- Specify that the resilient layer be dry when installed. Dampness affects its resilience.

- Build the floor in to the walls on all sides.

Site inspection
- Check the thickness of the concrete base and the mix for in-situ concrete, in particular the quantity and type of aggregate used.*

- Check all junctions with surrounding constructions, and ensure that all joints are well formed and free from honeycombing. Joints in beam/block and plank floors should be filled.*

- Ensure that the resilient quilt is dry on installation.

- *Most important:* undertake frequent and detailed inspections to ensure that there are no elements bridging between the floating layer and the base floor.*

- Ensure that the base floor is level and smooth before the resilient layer is laid.

- *Most important:* Check the grade and thickness of the resilient layer on site.*

- Ensure that the resilient layer is turned up at the edges to ensure that the edge of the floating layer does not touch the flanking construction.*

- Ensure that there is a narrow gap left between the bottom of the skirting board and the floating layer. Fill this gap only with a permanently-soft material.

- Ensure that the floor is built-in on all sides, if specified.*

*   Check with specification before inspecting.

Information sources
(See Appendix F)
BCA
BRE

Suspended timber-joist floors

Factors affecting performance

The AIRBORNE SOUND INSULATION of a suspended timber-joist floor is controlled mainly by the ISOLATION between ceiling and floor and their MASS, QUALITY OF CONSTRUCTION and FLANKING TRANSMISSION.

The IMPACT SOUND INSULATION of a suspended timber-joist floor is controlled mainly by the ISOLATION between the ceiling and the floor, their MASS, QUALITY OF CONSTRUCTION and FLANKING TRANSMISSION.

England and Wales, Approved Document E

The constructions which appear under this heading comply with Approved Document E, 1992 Edition, The Building Regulations 1991. It is not necessary to demonstrate that these constructions will meet the numerical performance standards associated with each of the Regulations. The mean performance figures are given only to assist the designer.

Construction A
Platform floor with absorbent material
(Figure 86(a))

Mass Floating layer either 18 mm tongue-and-groove timber or wood-based board, spot-bonded to 19 mm plasterboard, or two layers of cement bonded particle board, joints staggered, total thickness 24 mm. Floor base 12 mm timber boarding or wood-based board nailed to timber joists. Ceiling two layers of plasterboard with joints staggered; total thickness 30 mm

Isolation Resilient layer at least 25 mm mineral fibre, density 60 to 100 kg/m³. (The lower figure gives the better isolation but a 'softer' floor. In such cases additional support can be provided around the perimeter of the floor by a timber batten with a foam strip along the top attached to the wall.) Absorbent material 100 mm mineral-fibre quilt laid on the ceiling, density 10 kg/m³ or more

Quality Flooring material glued at all joints. No solid bridging between floating layer and base floor

Construction B
Ribbed floor with absorbent material
(Figure 86(b))

Mass Floating layer 18 mm tongue-and-groove timber or wood-based board, spot-bonded to plasterboard 19 mm thick, nailed or screwed to 45 mm × 45 mm battens located directly over the joists. Joists to be 45 mm wide. Ceiling two layers of plasterboard with joints staggered. Total thickness 30 mm

Isolation Resilient strips of mineral fibre at least 25 mm thick, density 80 to 140 kg/m³, laid on the joists. Absorbent material 100 mm unfaced mineral fibre laid on the ceiling, density 10 kg/m³ or more

Quality Flooring material glued at all joints. No solid bridging between floating floor and base floor

Construction C
Ribbed floor with heavy pugging
Figures 86 (c) and (d)

Mass Floating layer 18 mm tongue-and-groove timber or wood-based board, nailed to 45 mm × 45 mm battens located directly over the joists or between them. (For sheet materials, placing on joists is recommended.) Joists to be 45 mm wide. Ceiling either 19 mm dense plaster on expanded metal lath, or 6 mm plywood fixed under the joists plus two layers of plasterboard with joints staggered. Total thickness 25 mm. Pugging of mass 80 kg/m² laid on a polyethylene liner on the ceiling

Isolation Resilient strips of mineral fibre at least 25 mm thick, density 80 to 140 kg/m³, laid on the joists

Quality Flooring material glued at all joints. No solid bridging between floating floor and base floor

Pugging

The pugging between joists must provide a mass of at least 80 kg/m². It may be of the following types:

Traditional ash (75 mm),

2 mm to 10 mm limestone chips (60 mm),

2 mm to 10 mm whin aggregate (60 mm), or

dry sand (50 mm).

Figures in brackets show approximate thickness required to achieve 80 kg/m². (Other figures denote sieve size.) Sand should not be used in kitchens, bathrooms, shower rooms or water closet compartments where it may become wet and overload the ceiling.

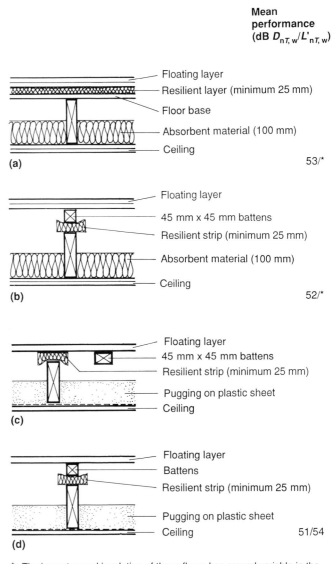

Mean performance (dB $D_{nT,w}/L'_{nT,w}$)

(a) — Floating layer / Resilient layer (minimum 25 mm) / Floor base / Absorbent material (100 mm) / Ceiling 53/*

(b) — Floating layer / 45 mm × 45 mm battens / Resilient strip (minimum 25 mm) / Absorbent material (100 mm) / Ceiling 52/*

(c) — Floating layer / 45 mm × 45 mm battens / Resilient strip (minimum 25 mm) / Pugging on plastic sheet / Ceiling

(d) — Floating layer / Battens / Resilient strip (minimum 25 mm) / Pugging on plastic sheet / Ceiling 51/54

* The impact sound insulation of these floors has proved variable in the field. If the advice on detailing is followed, the resulting impact insulation should be significantly better than the numerical values given on page 16.

Figure 86

Control of flanking transmission

Timber-frame walls

(Figure 87(a))

The gap between wall and floating layer should be filled using a resilient strip glued to the wall. A 3 mm gap should be left between skirting and floating layer. A seal is not necessary but if used it should be flexible.

Airpaths between the floor base and the wall should be blocked, including the space between joists.

Where joists are at right angles to the wall the junction between ceiling and wall lining should be sealed using tape or caulking.

Heavy masonry leaf

(Figure 87(b))

The mass of the masonry leaf (including any finish) should be at least 375 kg/m^2, both above and below the floor.

The gap between wall and floating layer should be filled with a resilient strip. A 3 mm gap should be left between skirting and floating layer. A seal is not necessary but if used it should be flexible.

Any normal method may be used to connect the floor base to the wall.

The junction between ceiling and wall-lining should be sealed using tape or caulking.

Lightweight masonry leaf

(Figure 87(c))

If the mass, including any plaster or plasterboard, is less than 375 kg/m^2 a freestanding panel should be used, as specified on page 44 for panel construction E or F.

The gap between panel and floating layer should be filled with a resilient strip. A 3 mm gap should be left between skirting and floating layer. A seal is not necessary but if used it should be flexible.

Any normal method may be used to connect the floor base to the wall, but airpaths between floor and wall cavities should be blocked.

The ceiling should be taken through to the masonry, and the junction with the freestanding panel should be sealed using tape or caulking.

Floor penetrations (excluding gas pipes)

(Figure 87(d))

Ducts or pipes penetrating a floor separating habitable rooms should be in an enclosure both above and below the floor. Openings in the floor should be no larger than necessary.

The material of the enclosure should have a mass of at least 15 kg/m^2. Either the enclosure should be lined, or the duct or pipe wrapped using 25 mm unfaced mineral fibre.

A nominal gap should be left between the enclosure and the floating layer, and a seal should be provided, using acrylic caulking or neoprene. The enclosure may go down to the floor base in the case of construction A (see page 60), but the enclosure should be isolated from the floating layer.

Penetrations of a separating floor by ducts and pipes should have fire protection in accordance with Approved Document B, Fire safety. The fire-stopping should be flexible, and should also prevent rigid contact between the pipe and the floor.

Floor penetrations for gas pipes

In the Gas Safety Regulations 1972 (SI. 1972/1189) and the Gas Safety (Installation and use) Regulations 1984 (SI. 1984/1358) there are requirements for the ventilation of ducts at each floor where they contain gas pipes. Gas pipes may be contained in a separate ventilated duct or they can remain unducted.

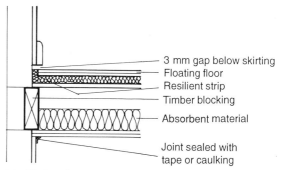

3 mm gap below skirting
Floating floor
Resilient strip
Timber blocking
Absorbent material

Joint sealed with tape or caulking

(a) Timber-frame wall with construction A (see page 60)

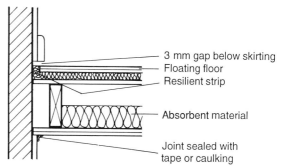

3 mm gap below skirting
Floating floor
Resilient strip

Absorbent material

Joint sealed with tape or caulking

(b) Heavy masonry leaf with construction A (see page 60)

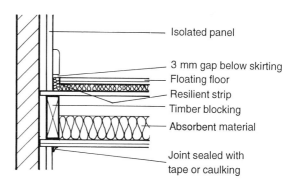

Isolated panel

3 mm gap below skirting
Floating floor
Resilient strip
Timber blocking
Absorbent material

Joint sealed with tape or caulking

(c) Lightweight masonry leaf with construction A (see page 60)

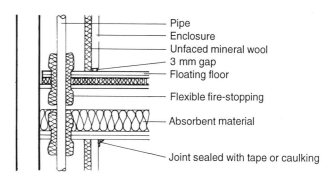

Pipe
Enclosure
Unfaced mineral wool
3 mm gap
Floating floor

Flexible fire-stopping

Absorbent material

Joint sealed with tape or caulking

(d) Floor penetrations (excluding gas pipes) with construction A (see page 60)

Figure 87

Northern Ireland, Technical Booklet G

The guidance under this heading complies with Technical Booklet G, June 1990, The Building Regulations (Northern Ireland) 1990. It is anticipated that future harmonisation with the Regulations for England and Wales will eliminate minor differences.

Suspended timber-joist floors

Floor constructions in Technical Booklet G are as shown on page 60 with the following exception:

> For constructions A and B (see Figures 86 (a) and (b)), the density of the absorbent material must be between 12 and 36 kg/m^3.

Control of flanking transmission

Specifications are as shown on page 61 with the following exceptions:

Heavy masonry leaf
The mass of any leaf (including any plaster) must be at least 355 kg/m^2 both above and below the floor.

Light masonry leaf
If the mass of any leaf (including any plaster) is less than 355 kg/m^2, a freestanding panel should be used, as specified on page 44, in panel constructions E and F.

Floor penetrations
Where a flue pipe penetrates the floor it shall be in a non-combustible enclosing duct with mineral-fibre absorbent lining.

Scotland, Part H

The guidance under this heading complies with Part H of the Building Standards (Scotland) Regulations 1991. At the time of writing, Part H is under review. It is anticipated that harmonisation with the Regulations for England and Wales will eliminate minor differences.

Suspended timber-joist floors

Floor constructions in Part H are as shown on page 60 with the following exceptions:

> This construction type is limited to use in buildings not more than four storeys high and whose fire resistance satisfies Part D.

> For constructions A and B (see Figures 86 (a) and (b)), the density of the absorbent material must be between 12 and 36 kg/m^3.

Part H offers a further suspended timber-joist floor construction, Type 4: Timber base with independent ceiling. It is intended primarily for use in refurbishment and flat conversions. Full details are given on page 71.

Control of flanking transmission

Specifications are as shown on page 61 with the following exceptions:

Heavy masonry leaf
The mass of any leaf (including any plaster) must be at least 355 kg/m^2 both above and below the floor.

Light masonry leaf
If the mass of any leaf (including any plaster) is less than 355 kg/m^2, a freestanding panel should be used, as specified on page 44, for panel construction E or F.

Floor penetrations
Penetrations of a separating wall by ducts and pipes must have fire protection in accordance with Part D.

Other suspended timber-joist separating floor constructions

Alternative constructions are permitted under the Regulations for England and Wales, Northern Ireland and Scotland but only if it can be demonstrated that a given numerical standard will be, or has been, achieved. The designer should make sure that any alternative construction will meet the numerical field or laboratory test requirements for the relevant country.

Detailed guidance is given on pages 16 to 19 .

In England and Wales and in Northern Ireland, test evidence must be obtained before construction. A trade organisation or manufacturer may be able to provide suitable information. The designer should ensure that the test results were obtained in circumstances as close as possible to those in the new application.

In Scotland, post-construction testing is acceptable.

Floors of lower performance

Reducing the MASS of the ceiling (and pugging) will result in reduced AIRBORNE and IMPACT SOUND INSULATION.

Removal or bridging of the resilient layer will cause a sharp reduction in the IMPACT SOUND INSULATION.

Floors of higher performance

FLANKING TRANSMISSION may prevent the attainment of significantly higher AIRBORNE and IMPACT SOUND INSULATION.

FLANKING TRANSMISSION can be controlled by installing an isolated wall lining, such as those specified on page 44, in front of the flanking walls.

For further IMPACT improvements, a soft floor covering with a good REDUCTION OF IMPACT SOUND LEVEL should be specified. Some practical examples are given on page 12.

Checklists
Design
● Specify the density and thickness of the mineral wool resilient layer. Check with the manufacturer that the product has long-term resilience under the concentrated loadings imposed by a domestic ribbed timber floor.

● Lightweight masonry flanking walls can cause flanking transmission problems unless an independent wall lining is adopted.

● Specify that pugging be dry, and be laid on plastic sheeting.

● Keep services to a minimum, and encase pipes.

● Specify that there be no bridging between the floating layer and the base.

● Check the specification of all junction details.

Site inspection
● *Most important:* ensure that the resilient layer on site is as specified.*

● Ensure that any pugging is dry and installed to the correct thickness on a plastic sheet.*

● Ensure that all layers in the construction are laid without gaps.

● Ensure that the resilient layer is turned up at the edges, or that a mineral-fibre or plastics foam strip fills the gap between the edge of the floating floor and the walls.*

● Ensure that there is a narrow gap between the floating floor and the skirting. Fill this gap only with a permanently-soft material.

● *Most important:* undertake frequent and detailed inspections to ensure that there are no nails nor any other elements, such as

services pipes and conduits, bridging between the floating floor and the base.*

* Check with specification before inspecting.

Information sources
(See Appendix F)
BG
BRE
TRADA

Walls in conversions

Factors affecting performance
The AIRBORNE SOUND INSULATION of an existing wall can be improved by increasing its MASS, improving the ISOLATION between its two faces or by introducing a new ISOLATED leaf. The extent of any improvement will be influenced by FLANKING TRANSMISSION and the QUALITY OF CONSTRUCTION of both the existing wall and any additional treatment.

Typical pre-conversion wall specifications
In general, there is no need to treat an existing wall which is generally similar to (for example within 15% of the mass of) one of the separating wall constructions which appear in Approved Document E for England and Wales (or Technical Booklet G for Northern Ireland).

Typical older house separating wall
(Figure 88)

Mass   215 mm brickwork, plastered both sides

This construction is similar to construction A on page 36 (Figure 54(a)). As long as its mass, including finishes, is not less than 319 kg/m^2, it will satisfy Approved Document E for England and Wales and Technical Booklet G for Northern Ireland without further treatment.

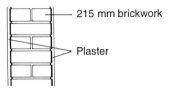

Figure 88

Masonry partitions
(Figure 89)

Mass   Masonry at least 100 mm thick, with plaster finish on both sides.

Where a partition of this type is to become a separating wall, Approved Document E for England and Wales and Technical Booklet G for Northern Ireland give details of a treatment which may be applied to one side of the partition only.

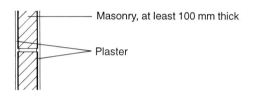

Figure 89

Other partitions
Where an existing masonry partition does not meet the minimum requirements, or the existing partition is of some other construction (for example a stud partition), the regulatory documents require the treatment described in the following paragraphs to be applied to both sides of the partition.

England and Wales, Approved Document E
The constructions which appear under this heading comply with Approved Document E, 1992 Edition, The Building Regulations 1991. It is not necessary to demonstrate that these constructions will meet a given performance standard.

Independent leaf and absorbent material
(Figure 90)

Mass   Existing wall requirements as above. Independent leaf thickness of each sheet 12.5 mm if a supporting framework is used, or total thickness of at least 30 mm if no framework is used

Isolation   Panel spaced 25 mm or more from the masonry core and fixed to floor and ceiling only. Framework spaced 13 mm or more from the masonry core. Absorber to be 25 mm mineral fibre, density 10 kg/m^3 or more

Quality   Joints between sheets staggered, and junctions with surrounding constructions sealed with tape or mastic. There should be no bridging of the cavity, except by the absorber, which should not be tightly compressed

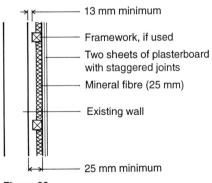

Figure 90

Typical junction details

Junction with a floating floor

(Figure 91)

The base of the independent leaf should be sealed, using tape or mastic. The gap between the independent leaf and the floating floor should be at least 10 mm wide, and filled using a resilient strip. A 3 mm gap should be left between skirting and floating floor. A seal is not necessary, but if used it should be flexible.

Junction with an existing or replacement ceiling

(Figure 91)

The ceiling should be taken through to the existing wall, and the junction with the independent leaf should be sealed, using tape or caulking.

Junction with an existing floor

(Figure 92)

The base of the independent leaf should be sealed, using tape or mastic.

Junction with an independent ceiling

(Figure 92)

The junction between the independent ceiling and the independent leaf should be sealed with tape or caulking.

The Approved Document does not explicitly require that the independent ceiling should not be connected to the existing wall, nor that the independent leaf should not be connected to the existing ceiling. Practical considerations will normally dictate that one or other of these elements should be carried through.

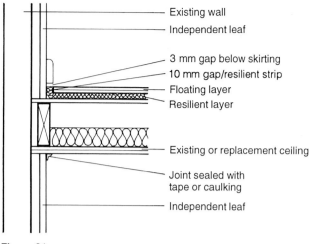

Existing wall
Independent leaf

3 mm gap below skirting
10 mm gap/resilient strip
Floating layer
Resilient layer

Existing or replacement ceiling

Joint sealed with tape or caulking

Independent leaf

Figure 91

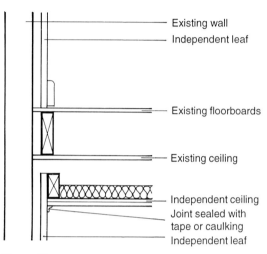

Existing wall
Independent leaf

Existing floorboards

Existing ceiling

Independent ceiling
Joint sealed with tape or caulking
Independent leaf

Figure 92

Northern Ireland, Technical Booklet G

It is anticipated that the Northern Ireland Regulations will be extended to cover flat conversions in early 1994. Technical Booklet G1 will contain deemed-to-satisfy provisions for sound insulation where an existing wall becomes a separating wall. The requirements are expected to be broadly as described for England and Wales.

Scotland, Part H

The guidance under this heading complies with Part H of the Building Standards (Scotland) Regulations 1991. At the time of writing, Part H is under review. It is anticipated that harmonisation with the Regulations for England and Wales will eliminate minor differences.

In Scotland, flat conversions are subject to the same sound-insulation requirements as new-build dwellings. The construction with independent leaf and absorbent material (see Figure 90) is not deemed-to-satisfy Part H. If it is used, it may be necessary to carry out sound-insulation testing after construction to demonstrate that it complies with new-build numerical performance standards, as described on page 19. In practice, the result would depend upon the construction of the existing wall and flanking conditions. New-build standards cannot be assured in all cases.

Other constructions for separating walls in conversions

Alternative treatments are permitted under the Regulations for England and Wales, Northern Ireland and Scotland, but only if it can be demonstrated that a given numerical standard will be, or has been, achieved. The designer should make sure that any alternative construction will meet the numerical field or laboratory test requirements associated with converted properties for the relevant country.

Detailed guidance is given on pages 16 to 19.

In England and Wales and in Northern Ireland, test evidence must be obtained before construction. A trade organisation or manufacturer may be able to provide suitable information. The designer should ensure that the test results were obtained in circumstances as close as possible to those in the new application.

In Scotland, post-construction testing is acceptable. New-build numerical performance standards must be achieved.

Treatment giving lower performance
Plasterboard on resilient channels
(Figure 93)

This construction can be used to improve the sound insulation of a timber-stud wall or partition where there is insufficient space for an independent leaf.

Mass	Existing plaster or plasterboard lining one side (other side removed). New plasterboard lining 32 mm or more in at least two layers
Isolation	Physical connections between new lining and existing studs via resilient bars only. Absorbent material 50 mm mineral fibre, density 10 kg/m^3 or more
Quality	Any gaps in the existing lining sealed. Any gaps between the joists of intermediate floors filled with plasterboard or timber blocking. Perimeter of new lining sealed with tape or caulking. Where possible, floorboards should not run through under the partition

Control of flanking transmission
Masonry walls

In many pre-conversion properties, existing masonry walls have the following common features:

● they are imperforate,

● they have been carried through all lightweight floors and partitions, and

● they are bonded into external flanking walls.

In these circumstances, flanking transmission is unlikely to prevent sound insulation up to new-build standards. For greater confidence of meeting this standard, the flanking details for solid masonry walls and separating walls given on page 37 should be adopted.

Timber-stud wall or partition

The following possible flanking paths around a timber-stud wall or partition may limit the improvement in sound insulation.

● Spaces between the joists of intermediate floors. These should be filled with plasterboard or timber blocks.

● Flanking timber-stud partitions. Where possible, the linings on the flanking partition should not be carried straight through.

● Flanking lightweight masonry walls (<360 kg/m^2, for example, half-brick walls). If an overall performance of around 48 dB $Dn_{T,w}$ or more is required, an independent leaf and absorbent material should be applied to the flanking wall.

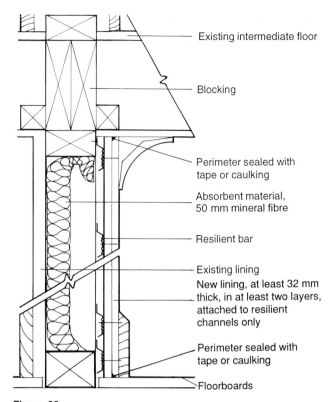

Existing intermediate floor

Blocking

Perimeter sealed with tape or caulking

Absorbent material, 50 mm mineral fibre

Resilient bar

Existing lining
New lining, at least 32 mm thick, in at least two layers, attached to resilient channels only

Perimeter sealed with tape or caulking

Floorboards

Figure 93

Checklists
Design

● Inspect the existing construction for gaps, particularly between the joists of intermediate floors, and specify that they be sealed up.

● Specify that there should be no physical contact between an independent lining and the existing construction.

● Specify good sealing around the perimeter of an independent lining.

Site inspection

● *Most important:* ensure there is no physical contact between an independent lining and the existing wall.

● *Most important:* where resilient bars are used, ensure that the only contact between the new plasterboard lining and the studs is via the resilient bars.

● Ensure that all layers in the construction are constructed without gaps and that there are no gaps in the existing layer.

● Ensure that any spaces between the joists of an intermediate floor are filled with plasterboard or timber blocks to a good tight fit.

Information sources

(See Appendix F)
BG
BRE
TRADA

Suspended timber-joist floors in conversions

Factors affecting performance

The AIRBORNE SOUND INSULATION of timber-joist floors is controlled by the ISOLATION between ceiling and floor, their MASS, QUALITY OF CONSTRUCTION and FLANKING TRANSMISSION.

In conversion work, the flanking walls are usually present already and may limit the attainable sound insulation. Check their construction and treat them as necessary to control FLANKING TRANSMISSION.

The IMPACT SOUND INSULATION of timber-joist floors is controlled by the ISOLATION between ceiling and floor, their MASS, QUALITY OF CONSTRUCTION and CONTROL OF FLANKING TRANSMISSION. Successful mechanical ISOLATION between floor and ceiling is of great importance to good IMPACT SOUND INSULATION.

Typical pre-conversion floor specifications

Typical older house floor
(Figure 94(a))

Mass   Floor 22 mm timber boards nailed to joists at 400-mm centres. Ceiling wood-lath and plaster

Typical Scottish tenement floor
(Figure 94(b))

Mass   Floor 22 mm timber boards nailed to joists at 400-mm centres. Pugging ('deafening') up to 150 mm ash on pugging boards. Ceiling wood-lath and plaster

In general, there is no need to replace an existing lath-and-plaster ceiling before treatment for sound insulation as long as it provides acceptable fire resistance. Where the ceiling must be removed, the replacement ceiling layer should comprise 30 mm plasterboard in two or three layers, with joints staggered and the perimeter sealed. If the wood-lath and plasterboard ceiling is removed and replaced with two thin sheets of plasterboard and no other action is taken, the sound insulation will be made much worse, typically around 38 dB $D_{\mathrm{n}T,\mathrm{w}}$ and 73 dB $L'_{\mathrm{n}T,\mathrm{w}}$.

In constructions where the existing floor boarding is to be retained as the top layer, any gaps should be sealed using caulking or by fixing hardboard on top.

England and Wales, Approved Document E

The constructions which appear under this heading comply with Approved Document E, 1992 Edition, The Building Regulations 1991. It is not necessary to demonstrate that these constructions will meet a given performance standard. The approximate performance figures are given only to assist the designer.

Floor treatment 1
Independent ceiling with absorbent material
(Figure 95)

Mass   Where the existing ceiling is not lath and plaster, it should comprise 30 mm plasterboard in two or three layers with joints staggered. A new independent ceiling should be at least 30 mm thick, comprising two layers of plasterboard with staggered joints

Isolation   A gap of 25 mm should be provided between the new joists and the existing ceiling. The cavity between the existing and the new ceiling layers should be at least 100 mm. It may be reduced in limited areas over window heads (see edge details in Figure 95). Absorber 100 mm mineral-fibre quilt above new ceiling, density 10 kg/m^3 or more

Quality   No gaps should be left in any layer. If there are gaps in the existing floor boarding they should be sealed, using caulking or by fixing hardboard on top. The perimeter of the independent ceiling should be sealed using tape or mastic

Approximate performance
(dB $D_{\mathrm{n}T,\,\mathrm{w}}/L'_{\mathrm{n}T,\,\mathrm{w}}$)

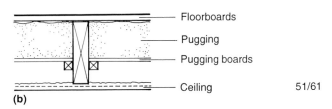

Floorboards

Joist

Ceiling   46 to 49/67 to 69

(a)

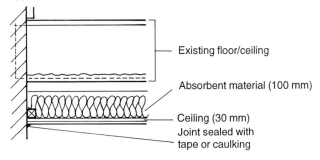

Floorboards

Pugging

Pugging boards

Ceiling   51/61

(b)

Figure 94

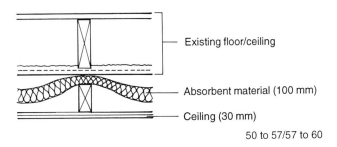

Existing floor/ceiling

Absorbent material (100 mm)

Ceiling (30 mm)

50 to 57/57 to 60

Typical edge detail

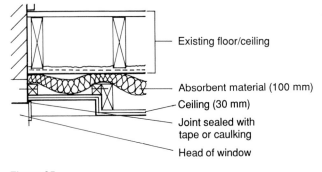

Existing floor/ceiling

Absorbent material (100 mm)

Ceiling (30 mm)
Joint sealed with tape or caulking

Edge detail where window head is close to the ceiling

Existing floor/ceiling

Absorbent material (100 mm)

Ceiling (30 mm)

Joint sealed with tape or caulking

Head of window

Figure 95

Floor treatment 2
Floating layer (platform floor)
(Figure 96)

Mass   Floating layer, either:

18 mm tongue-and-groove timber or wood-based board, spot-bonded to 19 mm plasterboard, or

a single or double layer of material having a mass of at least 25 kg/m²

If the existing floorboards are to be replaced, boarding at least 12 mm thick should be specified. Where the ceiling is not lath and plaster, it should comprise 30 mm plasterboard in two or three layers with joints staggered

Isolation   Resilient layer at least 25 mm mineral fibre, density 60 to 100 kg/m³. (The lower figure gives the better isolation but a 'softer' floor. In such cases additional support can be provided around the perimeter of the floor by a timber batten with a foam strip along the top attached to the wall.) Absorber, where possible, 100 mm mineral-fibre quilt laid between joists, density 10 kg/m³ or more

Quality   The flooring material should be glued at all joints. A 10 mm movement gap should be left around the floating layer and filled with resilient material. A 3 mm gap should be left between skirting and floating layer. A seal is not necessary but if used it should be flexible. The perimeter of any new ceiling should be sealed with tape or caulking

Floor treatment 3
Ribbed floor
The existing joists should be 45 mm wide. Additional strutting may be necessary between the joists to ensure stability of the floor after removal of the existing floor boarding.

With absorbent material
(Figure 97(a))

Mass   Floating layer, either:

18 mm tongue-and-groove timber or wood-based board, spot-bonded to plasterboard 19 mm thick, nailed or screwed to 45 mm × 45 mm battens located directly over the joists, or

a single or double layer having a mass of at least 25 kg/m², nailed or screwed to 45 mm × 45 mm battens located either directly over the joists or between them.

Where the existing ceiling is not lath and plaster, 30 mm plasterboard should be used in two or three layers with joints staggered

Isolation   Resilient strips of mineral fibre at least 25 mm thick, density 80 to 140 kg/m³, laid on the joists. Absorber 100 mm mineral-fibre quilt, density 10 kg/m³ or more

Quality   The flooring material should be glued at all joints. A movement gap 10 mm wide should be provided around the floating layer, and filled with resilient material. A 3 mm gap should be left between skirting and floating layer. A seal is not necessary but if used it should be flexible. The perimeter of any new ceiling should be sealed with tape or caulking. Nails should not be used to locate the battens during construction

With heavy pugging
(Figure 97(b))

Mass   Floating floor 18 mm tongue-and-groove timber or wood-based board, nailed or screwed to 45 mm × 45 mm battens located directly over the joists or between them. Ceiling as with absorbent material (see above). The capacity of the ceiling and joists to carry the load of the pugging (80 kg/m²) should be checked.

The following pugging materials may be used: Traditional ash (75 mm), or 2 mm to 10 mm limestone chips (60 mm), or 2 mm to 10 mm whin aggregate (60 mm), or dry sand (50 mm). Figures in brackets show approximate thickness required to achieve 80 kg/m², and other figures denote sieve size. Sand should not be used in kitchens, bathrooms, shower rooms or water closet compartments, where it may become wet and overload the ceiling.

Isolation and quality: As with absorbent material (see above)

Approximate performance
(dB $D_{nT,w}/L'_{nT,w}$)

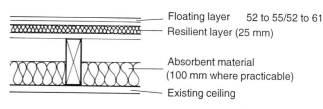

Floating layer   52 to 55/52 to 61
Resilient layer (25 mm)
Absorbent material (100 mm where practicable)
Existing ceiling

Typical edge detail

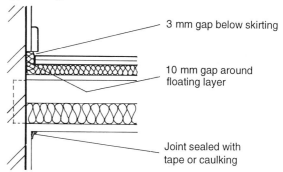

3 mm gap below skirting
10 mm gap around floating layer
Joint sealed with tape or caulking

Figure 96

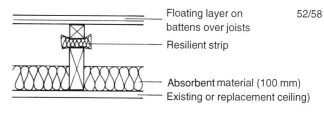

Floating layer on battens over joists   52/58
Resilient strip
Absorbent material (100 mm)
Existing or replacement ceiling)

Typical edge detail

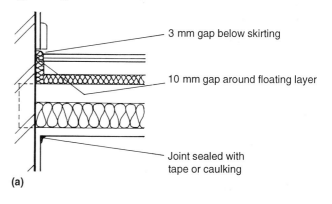

3 mm gap below skirting
10 mm gap around floating layer
Joint sealed with tape or caulking

(a)

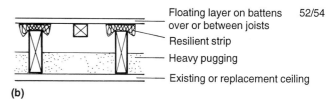

Floating layer on battens over or between joists   52/54
Resilient strip
Heavy pugging
Existing or replacement ceiling

(b)

Figure 97

Floor treatment 4   (Figure 98(a))
Alternative independent ceiling with absorbent material
(To be used only when a strong case can be made for not using floor treatments 1, 2 or 3.)

Mass	Ceiling 30 mm plasterboard in two or three layers with joints staggered
Isolation	The new ceiling should be supported on independent joists or suspended from the original joists by wire hangers not more than 2 mm in diameter or by metal straps not more than 25 mm × 0.5 mm. There should be not more than one fixing per square metre. The cavity between the floor and the new ceiling should be at least 100 mm. Absorber to be 100 mm mineral-fibre quilt above new ceiling, density 10 kg/m³
Quality	Any gaps in the existing floor boarding should be sealed, using caulking or by fixing hardboard on top. The perimeter of the independent ceiling should be sealed, using tape or mastic. Any strutting between the floor joists should not form a bridge between floor and ceiling

Floor treatment 5   (Figure 98(b))
Alternative floating layer (platform floor)
(To be used only when a strong case can be made for not using floor treatments 1, 2 or 3.)

Mass	Floating floor 18 mm tongue-and-groove boards or other boards with glued joints. Existing floorboards to be replaced, if necessary, with boarding at least 12 mm thick. Ceiling lath and plaster or plasterboard 30 mm thick in two or three layers with joints staggered, suspended from timber cross battens or suitable resilient hangers
Isolation	Resilient layer wood-fibre insulation board at least 13 mm thick. Refer to British Standard BS 1142: Specifications for fibre building boards. Part 3:1989 Insulation board (softboard). Absorber 50 mm mineral wool, density 10 kg/m³ or more. The absorber need be installed only if the floorboards or ceiling are to be removed for other reasons
Quality	The floating layer boarding should be glued at all joints. A movement gap of 10 mm should be left around the floating layer and filled with resilient material. The perimeter of any new ceiling should be sealed with tape or caulking

Stair treatment   (Figure 98(c))

Mass	Existing stair treads should receive a soft covering (for example carpet) at least 6 mm thick to control impact noise. An independent ceiling should be installed below the stair (see page 68). If there is a cupboard under the stair it should be lined with two layers of 12.5 mm plasterboard or material of similar mass; plasterboard at least 12.5 mm thick should be used to line the underside of the stair, and a small, heavy well fitted door should be installed in the cupboard
Isolation	Absorber 100 mm mineral-fibre quilt in cavity, density 10 kg/m³ or more

Piped services (other than gas)   (Figure 98(d))
Pipes and ducts that penetrate a floor separating habitable rooms in different dwellings should be enclosed above and below the floor. The material used to form the enclosure should have a mass of at least 15 kg/m². Either the enclosure should be lined or the pipe or duct within the enclosure should be wrapped with 25 mm of unfaced mineral wool. A 3 mm gap should be left between the enclosure and the floating layer of the floor, and the gap should be sealed with caulking or neoprene. The enclosure may go down to the floor base if floor treatment 2 is used (see page 69) but it should be isolated from the floating layer. Penetrations of a separating floor by ducts and pipes should have fire protection in accordance with Approved

Document B: Fire safety. The fire-stopping should be flexible and should also prevent rigid contact between the pipe and the floor.

Gas pipes
In the Gas Safety Regulations 1972 (SI. 1972/1178) and the Gas Safety (Installation and use) Regulations 1984 (SI. 1984/1358) there are requirements for the ventilation of ducts at each floor where they contain gas pipes. Gas pipes may be contained in a separate ventilated duct or they can remain unducted.

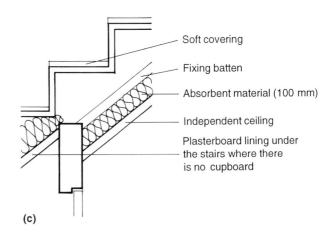

Approximate performance (dB $D_{nT,w}/L'_{nT,w}$)

(a)
- Existing floor — 47 to 56/63
- 100 mm minimum
- Absorbent material (100 mm)
- Ceiling on independent joists

(b)
- Floating layer — 49 to 51/64
- Resilient layer (13 mm wood-fibre insulation board)
- Absorbent material (50 mm)
- Existing or replacement ceiling

(c)
- Soft covering
- Fixing batten
- Absorbent material (100 mm)
- Independent ceiling
- Plasterboard lining under the stairs where there is no cupboard

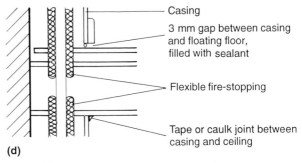

(d)
- Casing
- 3 mm gap between casing and floating floor, filled with sealant
- Flexible fire-stopping
- Tape or caulk joint between casing and ceiling

Figure 98

Northern Ireland, Technical Booklet G

It is anticipated that the Regulations will be extended to cover flat conversions in early 1994. Technical Booklet G1 will contain deemed-to-satisfy provisions for sound insulation where an existing floor becomes a separating floor. The requirements are expected to be broadly as for England and Wales, but floor treatments 4 and 5 and the stair treatment will not be included.

Scotland, Part H

Timber base with independent ceiling
(Figure 99(a))
In Scotland, flat conversions are subject to the same sound-insulation requirements as new-build dwellings.

This construction appears in Part H, and is suitable for use in converted properties. Its use is limited to buildings not more than four storeys high and with heavy masonry walls. Its fire resistance must satisfy part D of the Regulations

Mass   Floor 18 mm tongue-and-groove timber or wood-based boarding. Deafening on boards between joists 80 kg/m². Where the existing ceiling is not 19 mm dense plaster on lath, it should comprise 30 mm plasterboard in two layers, with joints staggered. Independent ceiling at least 30 mm thick, in two layers of plasterboard with staggered joints

Isolation   The cavity between the existing and the new ceiling layers should be at least 150 mm. Absorber to be 25 mm unfaced mineral fibre, density 12 to 36 kg/m³, draped over 45 mm thick joists

Quality   There should be no gaps in any layer. The edge of the independent ceiling should be sealed. There should be no solid contact between the existing and the independent ceilings

Control of flanking transmission
External wall or separating wall
The mass of leaf must be 355 kg/m² (including any plaster) both above and below the floor on at least three sides. The mass of the leaf on the fourth side must be at least 180 kg/m². Bearers should be used on walls to support the edges of the ceiling and to block airpaths. The junction of the ceiling and wall should be sealed using tape or caulking (Figure 99(b)).

Internal wall
The mass of masonry walls must be at least 180 kg/m². There are no restrictions for stud partitions. The edge of the ceiling should be supported and sealed as for external walls.

Piped services
Figure 99(c)
Ducts or pipes which penetrate the floor must be in an enclosure both above and below the floor. Either line the enclosure or wrap the duct or pipe within the enclosure with 25 mm unfaced mineral fibre. The material of the enclosure must have a mass of at least 15 kg/m². Where permitted by Part F, a flue pipe may penetrate the floor. The flue pipe must be in a non-combustible enclosing duct with mineral-fibre absorbent. The junction of ceiling and enclosure should be sealed using tape or caulking.

Checklists
Design
● Treat existing thin masonry walls to control flanking transmission. See page 61 for measures to control flanking transmission.

● Specify that pugging be dry and laid on plastic sheeting.

● Specify that there be no contact between an independent ceiling and the existing ceiling.

● Specify that there be no bridging of the resilient layer in any floating floor construction.

Approximate performance (dB $D_{nT,w}/L'_{nT,w}$)

52 to 57/57 to 60

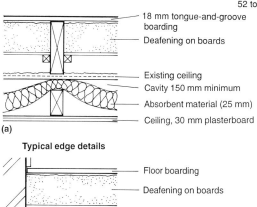

18 mm tongue-and-groove boarding
Deafening on boards

Existing ceiling
Cavity 150 mm minimum
Absorbent material (25 mm)
Ceiling, 30 mm plasterboard
(a)

Typical edge details

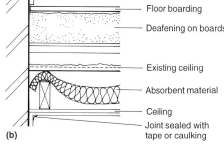

Floor boarding
Deafening on boards

Existing ceiling
Absorbent material

Ceiling
Joint sealed with tape or caulking
(b)

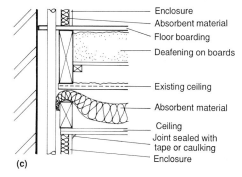

Enclosure
Absorbent material
Floor boarding
Deafening on boards

Existing ceiling

Absorbent material

Ceiling
Joint sealed with tape or caulking
Enclosure
(c)

Figure 99

● Check with the manufacturer of the resilient-layer material that the product has long-term resilience under the concentrated loadings imposed by a domestic ribbed timber floor.

● Specify sealing between the perimeter of the ceiling and surrounding constructions.

Site inspection
● *Most important:* ensure that there is no physical contact between an independent ceiling and the existing construction.*

● *Most important:* ensure that any resilient layer on site is as specified.*

● *Most important:* ensure that there is no bridging of any resilient layer in the construction, for example by nailing through a floating floor.*

● Ensure that any resilient layer is turned up at the edges and that there is a gap between any floating layer and the skirting.*

● Ensure that all layers in the construction are laid without gaps.

*   Check with job specification before inspecting.

Information sources
(See Appendix F)
BG
BRE
CIRIA
TRADA

Windows

Factors affecting performance

Single windows

The AIRBORNE SOUND INSULATION of a single-glazed window is controlled mainly by:

Mass   Thickness of the glass

Quality Detailing and construction to make window surrounds airtight

Sound-insulation values and further explanation of the physical principles are given on pages 30 and 31.

Double windows

The AIRBORNE SOUND INSULATION of secondary-glazed windows is controlled mainly by:

Isolation Cavity width and absorbent reveal linings

Mass   Pane thickness

Quality   Detailing and construction to make window surrounds airtight

Sound-insulation values and further explanation of the physical principles are given on pages 30 and 31.

Detailing

Always use compression seals in preference to brushes, if practicable:

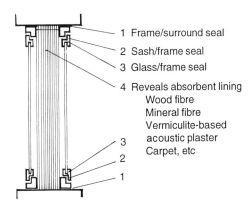

1 Frame/surround seal
2 Sash/frame seal
3 Glass/frame seal
4 Reveals absorbent lining
   Wood fibre
   Mineral fibre
   Vermiculite-based
   acoustic plaster
   Carpet, etc

3
2
1

Top-hung sash

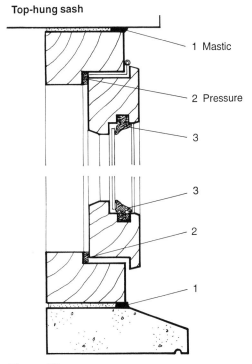

1 Mastic

2 Pressure

3

3

2

1

Figure 100

Pitfalls

Single windows

The most common pitfalls are:

● Failure to allow for ventilation. The effect of ventilation on the sound insulation of the building envelope is considered on pages 30 and 31

● Poor detailing or quality of construction leading to ineffective seals. Junctions between casements and frame and between frame and surround should be carefully detailed and implemented

Double windows

As for single glazing plus the following:

● Specification of inadequate cavity width causing a dip in the low-frequency performance due to the mass-air-mass resonance

● Selection of identical pane thicknesses leading to an identical CRITICAL FREQUENCY DIP in both panels. This usually occurs at high frequencies so is likely to be a problem only with a high-frequency source.

Sliding sashes – brushes

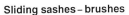
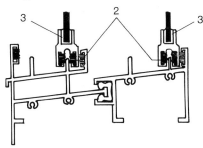

3   2   3

Centre-hung sash

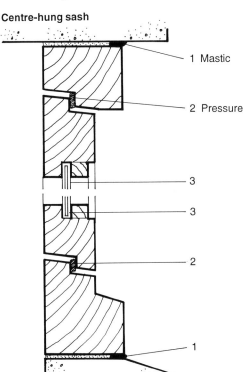

1 Mastic

2 Pressure

3

3

2

1

Doors

Factors affecting performance

The AIRBORNE SOUND INSULATION of a door depends on:

Mass   Thickness and density of the door leaf

Quality   Detailing and construction to ensure an airtight seal between the door and its frame and between the frame and its surround

Sound-insulation values and further explanation of the physical principles are given on pages 30 and 31.

Where sound insulation in excess of 36 dB R_w or 30 dB $R_{A(traffic)}$ is required, a lobby should be introduced. For best results, sound-absorbing finishes should be installed in the lobby (for example carpet and absorbent ceiling tiles).

The noise problem caused by slamming doors can be alleviated by fitting a closer, designed to reduce the speed of impact while pulling the door shut.

Pitfalls

The most common pitfalls are:

● Specification of inadequate ironmongery to compress the seals effectively

● Gaps in the sealing

● Doors not hung sufficiently accurately to compress the seals reliably

● Gaps between the frame and the surrounding wall

● Apertures through the door leaf, such as a key hole, letter box or cat flap

Where an existing door is to be improved, it may be necessary to enlarge the rebates and, therefore, relocate the door furniture.

Detailing (Figure 101)

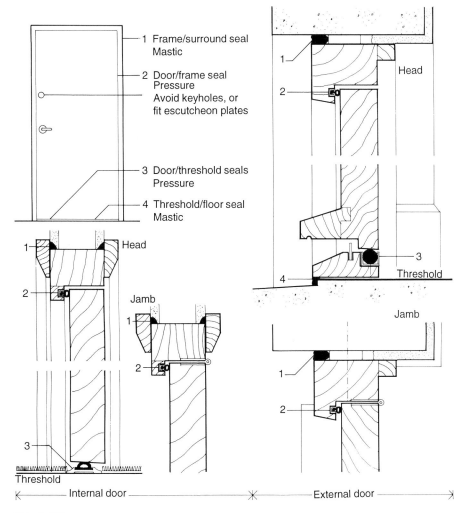

1 Frame/surround seal
  Mastic

2 Door/frame seal
  Pressure
  Avoid keyholes, or
  fit escutcheon plates

3 Door/threshold seals
  Pressure

4 Threshold/floor seal
  Mastic

Head

Jamb

Threshold

Head

Threshold

Jamb

Internal door — External door

Figure 101

Pitched roofs

Factors affecting performance

The AIRBORNE SOUND INSULATION of a pitched roof depends mainly upon:

Mass Of roof and ceiling

Isolation Between roof and ceiling, which depends upon the air gap and the presence of absorption in the roof space

Quality Absence of gaps, as far as ventilation requirements allow

Sound-insulation values and further explanation of the physical principles are given on pages 30 and 31.

Detailing (Figure 102)

Note

Where $R_{A(traffic)}$ of 40 dB or more is required, seek expert advice

Pitfalls

The most common pitfalls are:

● Inadequate ceiling mass.

● Excessive gaps in the roof. (This conflicts with the normal ventilation requirements for roof spaces. For this reason, the details above are restricted to improvements to the ceiling layer. They will still permit eaves ventilation. Boarding under the battens can be effective if the ventilation problem can be overcome).

● Poorly designed trap door. The trap door should be of similar mass and construction to the ceiling layer. To bring a timber trap door up to this standard, plasterboard can be fixed to the back of the trap door and compression seals fitted, similar to those used for doors (see page 73).

● Poorly-detailed flue penetrations.

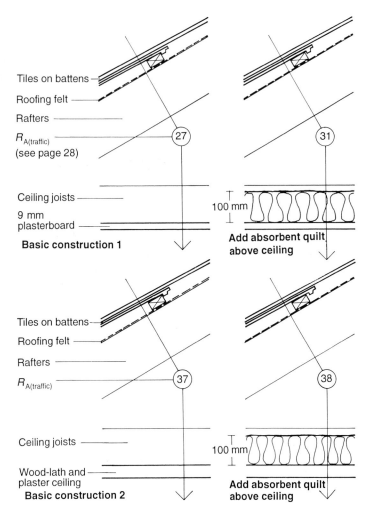

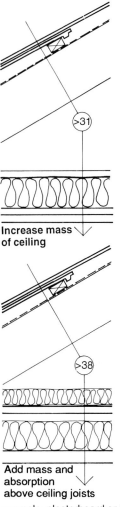

Figure 102

Flat roofs

Factors affecting performance

Concrete roof

The AIRBORNE SOUND INSULATION of a flat concrete roof depends mainly upon:

Mass   Density and thickness of concrete

Quality   Detailing and construction to ensure an absence of gaps

Sound-insulation values and further explanation of the physical principles are given on pages 30 and 31.

Timber-joist roof

The AIRBORNE SOUND INSULATION of a timber-joist roof depends mainly upon:

Mass   Density and thickness of roof deck and ceiling

Isolation Airgap and absorbent material in the cavity

Quality   Absence of gaps, as far as ventilation requirements allow

Sound-insulation values and further explanation of the physical principles are given on pages 30 and 31.

Pitfalls

Concrete roof

The main pitfall is the occurrence of unexpected gaps caused by services penetrations, etc.

There should be no bridging between a floating walking surface and the roof slab.

Timber-joist roof

The main pitfall is the provision of inadequate ceiling mass.

Detailing (Figure 103)

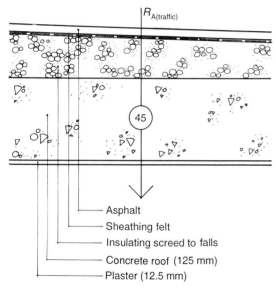

- Asphalt
- Sheathing felt
- Insulating screed to falls
- Concrete roof (125 mm)
- Plaster (12.5 mm)

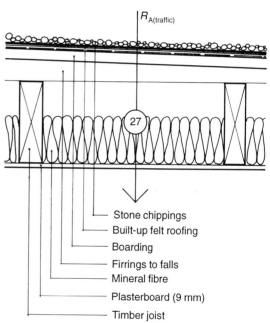

- Stone chippings
- Built-up felt roofing
- Boarding
- Firrings to falls
- Mineral fibre
- Plasterboard (9 mm)
- Timber joist

For $R_{A(traffic)}$ >27 dB, increase the mass of the roof and/or ceiling, or install an independent ceiling (see page 64)

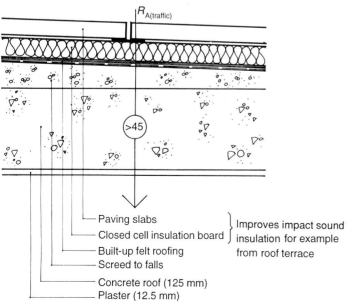

- Paving slabs   ⎱ Improves impact sound
- Closed cell insulation board ⎰ insulation for example
- Built-up felt roofing   from roof terrace
- Screed to falls
- Concrete roof (125 mm)
- Plaster (12.5 mm)

Figure 103

Part D
Worked examples

Site noise assessment

The first example illustrates techniques used to plan a site against external noise, and to design the building envelope of new dwellings affected by external noise. The example describes a site for new housing which is subject to noise from road traffic, a railway line, a factory and a dairy. Each source is considered separately in terms of the following:

● Noise criteria

● Measurement and calculation of site noise levels

● Calculation of the sound-insulation requirements of the building envelope

Details of the necessary calculation methods are given in Appendix A, pages 111 to 120.

Design of new dwellings

Four examples are presented, to illustrate how to control internal noise in new housing using the methods given in Parts A to C. The examples deal with the following aspects of design:

● Planning to control internal noise

● How to meet the requirements of the Building Regulations

● Design and construction of elements to meet numerical sound-insulation criteria

● Assessment of noise transmission through a wall

The examples cover the design of masonry, timber-frame and high-rise dwellings, and the design of a quiet room.

Conversion properties

Examples of two types of dwelling conversion:

A typical older, two-storey terraced property
The main planning considerations are covered and suitable conversion plans proposed. Construction methods are given to meet current Building Regulations.

Loft conversions
This page deals with the sound insulation of the roof and separating wall in three practical cases. Separating wall construction methods are given to meet current Building Regulations.

Noise problems in existing dwellings

Thirteen self-contained examples of noise problems which arise in existing dwellings are described here. They fall into five main groups:

● *Outside-to-inside sound insulation*
  Road traffic, aircraft and other external sources

● *Separating element sound insulation*
  Masonry and lightweight separating walls, concrete and timber-joist separating floors

● *Specially insulated rooms*
  Quiet room and music room

● *Noise nuisance and the law*
  Legal remedies to noise problems

● *Domestic mechanical noise sources*
  Domestic machine, water pump and lift

Each example gives a typical brief followed by a technical solution to the problem, including details of any calculations necessary to solve the problem.

Site noise assessment

Brief

Development is to take place on an infill site in a mixed residential and light industrial area. To the north of the site is a dairy; to the east is a food processing factory; to the south are a railway and an elevated trunk road, and to the west there is existing housing. An appraisal is to be made of the noise-on-site arising from each source separately.

The pre-development site plan is shown in Figure 104. The dwellings are two-storey houses and three-storey flats. The calculation methods used are illustrated by detailed examples in Appendix A. The numbered plots appear in these examples. They have been selected because they best illustrate the methods given in parts A and B of this manual. The results of this detailed appraisal are summarised here.

Noise criteria

The local authority planners have set the following maximum design noise levels in their planning conditions:

In any dwelling	45 dB $L_{Aeq,16h}$ (7 am to 11 pm)
	35 dB $L_{Aeq,8h}$ (11 pm to 7 am)
Any private garden	65 dB $L_{Aeq,0.5h}$ (at any time)

Noise assessment

Road-traffic noise levels on site can be calculated on the basis of vehicle counts and site drawings. A site noise survey is also necessary, however, as train noise and industrial sources cannot be predicted accurately.

Road traffic noise

(See Appendix A, page 112)

The Highways Department of the local authority has supplied predicted traffic flows on the nearby trunk road for the next 15 years. These data are used with Calculation of road traffic noise by DOE and the Welsh Office, HMSO, 1988, to find the position of the 68 dB $L_{A10,18h}$ contour on site and to obtain the noise level at the position of the nearest exposed house. The results are compared here with the criteria set by the local authority and with those suggested on page 20.

Results

The 68 dB $L_{A10,18h}$ noise contour is 67 m from the road. Much of the site exceeds this facade level which corresponds to approximately 66 dB $L_{Aeq,16h}$. This is close to the local authority's external criterion of 65 dB $L_{Aeq,0.5h}$, so measures should be adopted on this site to control traffic noise.

The house on plot 80 receives a level of 61 dB $L_{Aeq,18h}$ (approximately 59 dB $L_{Aeq,16h}$) at ground-floor level, and 63 dB $L_{A10,18h}$ (approximately 61 dB $L_{Aeq,16h}$) at first-floor level. These levels must be reduced by approximately 14 dB and 16 dB, respectively, to meet the local authority's daytime internal criterion of 45 dB $L_{Aeq,16h}$ inside the dwelling. These requirements are comfortably met by 4 mm single glazing.

The garden behind the house on plot 80 receives a level of 45 dB $L_{A10,18h}$ (approximately 43 dB $L_{Aeq,16h}$). This will meet the local authority's criterion, which is expressed in terms of $L_{Aeq,0.5h}$, as the variation in level over the 16-hour day is unlikely to be sufficient for a value of 65 dB to be exceeded during the noisiest half-hour.

On more severely-exposed sites it may be necessary to consider the use of barriers and barrier blocks for noise control (see page 22).

Railway noise

(See Appendix A, page 114)

The main stages in the railway noise assessment are:

● Identification of the different train types which use the track

● Measurement of the sound exposure level associated with each train type at a reference position on-site close to the track

● Determination of the number of movements of each train type which uses the track

● Combination of level data and number data, to obtain the equivalent continuous sound-pressure level for the measurement position

● If possible, measurement of sound levels at various distances from the track, to assess the reduction of level with distance

● Comparison of the results with the criteria set by the local authority and with those suggested on page 20

Results

Calculations based on noise measurements give the following noise levels 12 m from the track:

	$L_{Aeq,T}$ (dB)
Day (7 am to 11 pm)	65
Night (11 pm to 7 am)	59
Busiest half-hour	67

On the undeveloped site, the reduction in level across the site cannot be determined by noise measurement because there are existing buildings close to the southern boundary. If the site is grassed, levels fall off at a rate of about 5 dB per doubling of distance from the track.

On this basis, the following noise levels have been predicted at the facade of plot 80:

Daytime, external	51	dB $L_{Aeq,16h}$
Criterion, internal	45	dB $L_{Aeq,16h}$
Sound insulation needed	6	dB
Night-time level, external	45	dB $L_{Aeq,16h}$
Criterion, internal	35	dB $L_{Aeq,16h}$
Sound insulation needed	10	dB

The sound-insulation requirements can be met by partly-open windows.

The garden criterion will be met at locations more than 16 m from the track.

Factory noise

(See Appendix A, page 116)

The factory to the east of the site operates 24 hours a day. It is a tall building, with noisy plant situated on the roof. Noise levels on site do not fall off uniformly with distance. As the distance from the factory increases, the roof-top plant comes more into view and the barrier effect of the edge of the roof is progressively lost. Site noise measurements are made to determine the sound distribution, and the results are used to assess conditions at plots 30 and 52 in terms of the local authority's criteria.

The maximum continuous noise level measured on site is 50 dB L_{Aeq}. This is well below the criterion for gardens of 65 dB $L_{Aeq,0.5h}$. The external level at plot 30 is 48 dB L_{Aeq}. To prevent factory noise exceeding 35 dB $L_{Aeq,8h}$ inside the dwelling at night, it is necessary to restrict the window opening area to approximately 0.125 m².

Plot 52 is partially shielded by the house on plot 30. Consequently, its external noise level is only 44 dB L_{Aeq}.

Dairy noise

(See Appendix A, page 118)

Noise is created by the dairy between 5 am and 8 am. The tractor units of delivery lorries start up, reverse into their loaded trailers and depart, while milk floats are continuously loaded at a nearby loading bay. Noise measurements are carried out for each activity, and the results are used to assess conditions for nearby dwellings in terms of the noise criteria set by the local authority.

The main stages in the assessment are:

● Identification of the individual activities which contribute to the noise levels heard on site

● Measurement of noise levels (short-term L_{Aeq}) for each activity, recording the distance from the microphone, the duration of each event, and the number of occurrences

● Conversion of the equivalent continuous sound-pressure level (L_{Aeq}) value for each activity to a sound exposure level (L_{AE})

● Adjustment of the L_{AE} values for the distance to the plot under consideration

● Combination of the L_{AE} values to obtain the equivalent continuous sound-pressure level

Results

Calculations based on noise measurements give the following levels in the garden of plot 28:

No barrier	77 dB $L_{Aeq,0.5h}$
With 2 m high boundary fence	69 dB $L_{Aeq,0.5h}$
Criterion	65 dB $L_{Aeq,0.5h}$

Recommendation

Noise from the dairy exceeds the criterion for gardens set by the local authority. Figure 105 shows a possible alternative site layout which removes private gardens from the site boundary near to the dairy. The low-rise housing to the north of the site is replaced by high-rise blocks which do not have private gardens. The night-time criterion of 35 dB $L_{Aeq,8h}$ internally is just achieved, in the high-rise housing, by closed 4 mm single-glazing in rooms which overlook the dairy.

As the windows cannot be opened for ventilation, it will be necessary to provide a sound-attenuated ventilation system. Unit ventilators meeting the requirements of the Noise Insulation Regulations (see page 30) would meet the acoustic requirements.

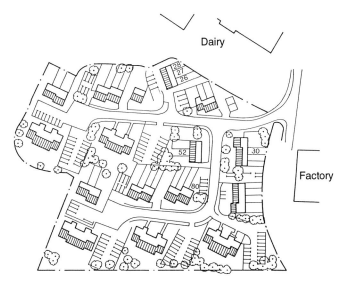

Road and railway

Figure 104

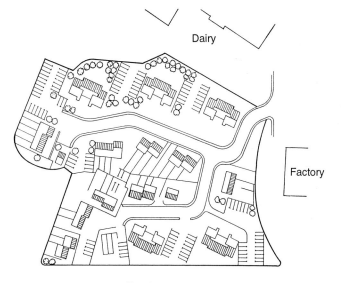

Road and railway

Figure 105

Design of new dwellings

Design of masonry dwellings
Brief
Figure 106 shows the layout proposed for a two-storey block of flats of masonry construction. For reasons other than sound insulation, the client prefers to adopt the following forms of construction:

Separating and flanking external walls: cavity concrete blockwork with a dry lining on plaster dabs

Separating floor: Concrete beam and block with a floating screed

Load-bearing partitions: 115 mm thick autoclaved aerated concrete blockwork: 9.5 mm plasterboard lining each side

Non load-bearing partitions: 75 mm timber stud, 12.5 mm plasterboard each side

● Comment on the planning, in terms of the control of internal noise, and make recommendations for planning improvements.

● Comment on the choice of separating elements, and make recommendations to ensure that reasonable sound insulation will be achieved.

● How much sound insulation is required from internal partitions? Do the proposed forms of construction meet these requirements?

Planning to control internal noise
The dwelling plans are handed and stacked to ensure that the rooms of adjacent dwellings are compatible (see page 25). The domestic water services have been poorly planned in relation to sensitive rooms (see page 31):

● The soil stack and general bathroom plumbing are located on a wall next to the bedroom.

● The hot water cylinder is in a bedroom cupboard.

● Sink and washing machine are fixed to the bedroom partition.

Recommendations

● Soil stacks should be routed away from sensitive rooms.

● Water services should be zoned away from sensitive rooms.

● Access panels to services should be outside sensitive rooms.

Figure 107 shows an improved layout. The bath, basin, WC, linen cupboard, soil stack and all associated plumbing have been moved away from the bedroom and fixed to an external masonry wall.

Design of separating elements and partitions
Separating walls
The separating walls should be identified using the guidance given on page 26. There are two walls in this example (Figure 108), the main separating wall between dwellings, and the separating wall between ground-floor living room and the staircase which serves the first-floor flat.

Approved Document E for England and Wales gives one construction which meets client requirements: the cavity lightweight aggregate concrete blockwork separating wall (see construction C on page 40 and Figure 109). This construction does not appear in the Regulations for Northern Ireland or Scotland. (However, a dry-lined construction is available for use only where a step and/or stagger of at least 300 mm is provided. See pages 40 to 42.)

The cavity lightweight aggregate concrete blockwork wall gives a mean field performance of 52 dB $D_{nT,w}$. As this just meets the minimum numerical requirements associated with the various building regulations, special care should be taken in the detailing and workmanship to ensure that reasonable sound insulation is achieved in practice.

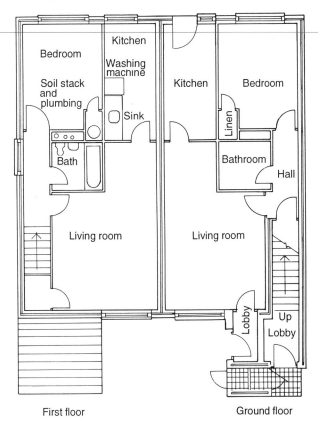

First floor   Ground floor

Figure 106

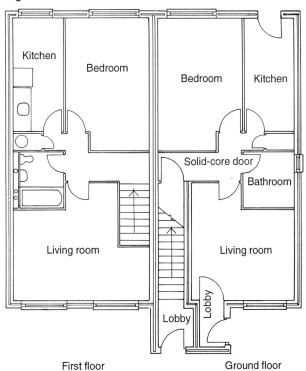

First floor   Ground floor

Figure 107

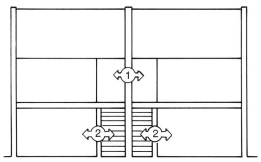

Figure 108

The separating wall should have a mass of at least 300 kg/m². Flanking transmission should be controlled by specifying the inner leaf of the external flanking wall with a mass of at least 120 kg/m². The same separating wall specification should be used for the wall between the ground-floor flat and the first-floor access stairs.

The design checklist given on page 43 should be used to ensure that all necessary precautions have been taken in the design of the wall. The separating wall should be inspected frequently during construction, and reference should be made to the site inspection checklist given on page 43.

Separating floor

Floor base C on page 56 appears in the regulatory documents for England and Wales, Northern Ireland and Scotland, and meets client requirements (see Figure 110). The airborne sound insulation associated with this floor is 54 dB $D_{nT,w}$. If care is taken in detailing and workmanship, it can be expected to provide reasonable sound insulation.

The recommendations given on page 57 for the control of flanking transmission should be followed. The design checklist given on page 59 should be used to ensure that all precautions have been taken in the design and detailing of the floor. The separating floor should be inspected frequently during construction, and reference should be made to the site inspection checklist given on page 59.

Access stairs

It may not be practicable to construct a concrete staircase giving similar mass and isolation to the separating floor construction. A carpet should be specified to reduce impact noise, and a well-fitting solid-core door should be installed in the cupboard under the stairs to improve the airborne and impact sound insulation of the staircase. These specifications should be cleared with the building control officer, as they do not appear in the regulatory documents.

Partitions

There are no regulatory requirements for the sound insulation of partitions within dwellings. The standard suggested in this manual (page 27) is that general partitions should be designed to give at least 38 dB $D_{nT,w}$.

The client's preferred constructions give the following sound-insulation values:

75 mm timber stud, 12.5 mm plasterboard each
side (see page 51)   36 dB R_w

115 mm aerated autoclaved concrete blockwork,
9.5 mm plasterboard dry lining (see page 37)   47 dB R_w

The sound insulation of the masonry partition exceeds the criterion. The sound insulation of the timber-stud partition appears to be a little low.

Strictly, these values should be converted from R_w to $D_{nT,w}$ value by taking into consideration the surface area of the partition and the volume of the receiving room. See page 13 and examples on pages 90 and 101

According to Table 16, page 51, the sound insulation of the timber-stud partition can be improved from 36 to 40 dB R_w by including a 25 mm absorbent quilt in the cavity (see Figure 111).

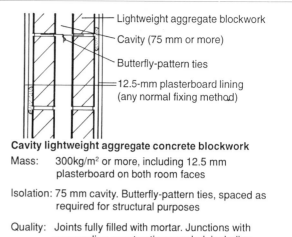

Cavity lightweight aggregate concrete blockwork

Mass: 300kg/m² or more, including 12.5 mm plasterboard on both room faces

Isolation: 75 mm cavity. Butterfly-pattern ties, spaced as required for structural purposes

Quality: Joints fully filled with mortar. Junctions with surrounding constructions sealed, including those behind linings

Figure 109

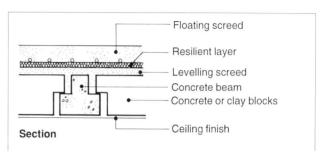

Section

Screed

Mass: 55 mm cement sand screed with 20 mm to 50 mm wire mesh (to protect the resilient layer while the screed is being laid)

Quality: Screed must not enter or penetrate the resilient layer. The resilient layer may be protected by a paper facing

Resilient layer

Isolation: Mineral wool, density 36 kg/m³, thickness at least 25 mm

Quality: Mineral wool tightly butted and turned up at the edges at the floor perimeter. The mineral wool should be paper-faced on the upper side to prevent the screed entering the layer

Concrete beams with infilling blocks

Mass: 300 kg/m² including levelling screed and ceiling finish

Quality: The finish should be level. Joints should be filled to avoid airpaths

Figure 110

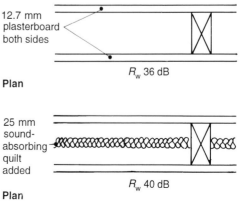

Figure 111

Design of timber-frame dwellings

Brief 1

Figure 112 shows the layout proposed for two-storey flats which are to be built in timber-frame construction. The contractor wishes to build from prefabricated panels, using a crane to assist with erection on site.

● Comment on the choice of separating elements, and make recommendations to ensure that reasonable sound insulation will be achieved.

● How much sound insulation is required from internal partitions? What forms of construction can be used to meet these requirements?

Design of separating elements and partitions

Separating walls

The separating walls should be identified using the guidance given on page 26. There are two separating walls in this example (Figure 113):

the main separating wall between dwellings, and

the separating wall between ground-floor living room and first-floor access staircase.

The timber-frame construction (Figure 72) on page 48 meets the requirements of the brief and complies with the regulatory documents for England and Wales, Northern Ireland and Scotland, subject to minor differences (see page 50).

The position and thickness of the mineral-fibre absorbing quilt can vary. If a single 25 mm quilt is used, more site control is necessary to ensure that it is correctly positioned between the frames. Therefore, for a crane-handled building system, a 50 mm quilt fixed to one frame (see Figure 114) or a 25 mm quilt, fixed to both frames, is more appropriate. The same wall specification should be used for the separating wall between the ground-floor flat and the first-floor access stairs.

The recommendations given on page 51 for the control of flanking transmission should be followed. The design checklist, also on page 51, should be used to ensure that all necessary precautions have been taken in the design of the wall. The separating wall should be inspected frequently during construction, and reference should be made to the site inspection checklist on page 51.

Separating floors

The floor specifications which appear in Approved Document E for England and Wales are given on page 60. There are two types of floating floor to choose from:

● Platform floor, where the resilient layer is supported on a continuous timber floor-base on top of the joists (see Figure 115)

● Ribbed floor, where the resilient layer is supported directly on top of the joists

The platform floor is more appropriate for use with a crane-handled system, because the floor base which supports the resilient layer can be designed to provide racking strength, to prevent collapse of the panel during handling. It also provides an immediate working platform for the erecting crew.

The recommendations given on page 61 for the control of flanking transmission should be followed. The design checklist given on page 63 should be used to ensure that all precautions have been taken in the design and detailing of the floor. The separating floor should be inspected frequently during construction, and reference should be made to the site inspection checklist on page 63.

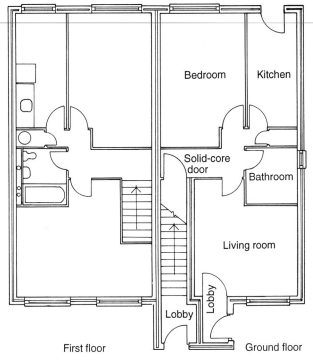

Figure 112

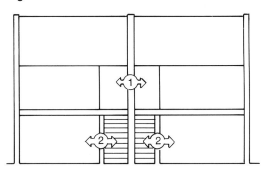

Figure 113

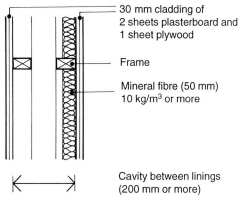

30 mm cladding of 2 sheets plasterboard and 1 sheet plywood

Frame

Mineral fibre (50 mm) 10 kg/m^3 or more

Cavity between linings (200 mm or more)

Separating wall
Figure 114

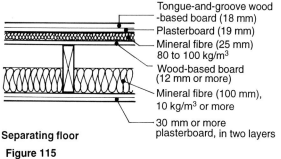

Tongue-and-groove wood-based board (18 mm)
Plasterboard (19 mm)
Mineral fibre (25 mm) 80 to 100 kg/m^3
Wood-based board (12 mm)
Mineral fibre (100 mm), 10 kg/m^3 or more
30 mm or more plasterboard, in two layers

Separating floor
Figure 115

Access stairs
(Figure 116)

It may not be practicable to construct a staircase giving similar mass and isolation to that of the separating floor construction. Airborne and impact sound can be controlled by adopting the measures proposed in Approved Document E for stairs in conversions, where there is a cupboard under the stairs (see page 70).

- Stair treads: soft covering (carpet, for example) at least 6 mm thick

- Ceiling: at least one skin of 12.5 mm plasterboard to the underside of the stairs

- Cupboard lining: at least two layers of 12.5 mm plasterboard or material of similar mass

- Cupboard door: small, heavy, well-fitted

These specifications should be cleared in advance with the building control officer or approved certifier, as they are intended for use in conversions rather than in new-build dwellings.

Partitions

There are no regulatory requirements for the sound insulation of partitions within dwellings. The standard suggested in this manual (page 27) is that general partitions be designed to give at least 38 dB $D_{nT,w}$.

The following timber-frame construction gives 40 dB R_w (Figure 117). See also Table 16 on page 51):

> 75 mm timber stud, 12.5 mm plasterboard each side, 25 mm absorbent quilt in the cavity

Strictly, this value should be converted from R_w to a $D_{nT,w}$ value by taking into consideration the surface area of the partition and the volume of the receiving room. See page 13 and examples on pages 90 and 101.

The partition should be constructed without gaps, and sealant should be applied around the edges.

Brief 2

During the design of a timber-frame house, the client insists on electrical sockets on separating walls. What precautions should be taken to ensure the minimum loss of sound insulation?

Recommendations

Holes or gaps which lead to a direct airpath through a wall may seriously affect its sound insulation (see page 15). If possible, services penetrations through timber-frame separating walls should be avoided. (The regulatory documents for Northern Ireland and Scotland do not permit services to be contained in the wall.)

Socket outlets should either be located on other walls, or surface trunking or skirting ducts should be used. If surface trunking is unacceptable, electrical socket outlets which are set in the cladding of the separating wall should have a similar thickness of cladding enclosing the socket box. Back-to-back socket outlets should be avoided.

Figure 118 shows a suitable detail for use in situations where flush-mounted electrical socket outlets are unavoidable.

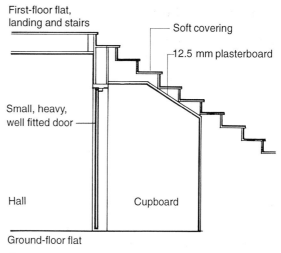

Figure 116

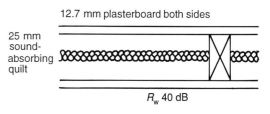

Plan

Figure 117

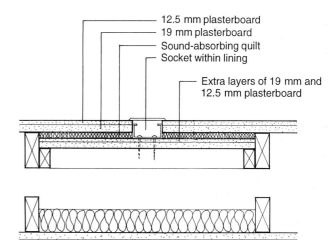

Figure 118

Design of high-rise dwellings

Brief

Figure 119 shows the layout proposed for a high-rise block of flats. There are four flats of type A on each floor except the ground floor, where there are two type-A flats and two type-B flats. The ground-floor plan differs from the rest because space has been provided for a pump room.

- Comment on the planning, with regard to the control of internal noise.

- Calculate the noise level in flat B resulting from the neighbouring pump room.

- Make recommendations to eliminate any potential noise problems.

Data required

- Sound-level measurements have been made in a pump room which houses similar plant. The results are tabulated on page 85.

- The pumps will sit on correctly designed anti-vibration mountings (see page 33).

- The wall between the flat and the pump room is brickwork, mass 480 kg/m², plastered each side.

- The absorption in the adjacent bedroom can be assumed to correspond to 10 m².

Planning to control internal noise

First floor and above

- The dwelling plans are handed and stacked to ensure that the rooms of adjacent dwellings are compatible (see page 25).

- Kitchens and bathrooms are located next to the lift shaft or stairs, and therefore provide a buffer zone between sensitive rooms and the lift shaft.

- Bedrooms are located well away from the lifts and well away from the bathroom service duct (see page 33).

- Less sensitive rooms are located next to separating walls.

Therefore the layout of flats on the first floor and above is satisfactory.

Ground floor

The planning of this floor gives rise to two potential noise problems for the smaller bedroom in the ground-floor flat:

- The bedroom in flat type B is located beneath the living room and bathroom of first-floor flat type A.

- The bedroom shares a wall with the pump room, which is a potential source of disturbance.

Calculations indicate that in the bedroom the noise level attributable to pump noise is 38 dB(A). Details of the calculation are given on page 85. The calculated level should be compared with the criteria for mechanical services given on page 20. A level of up to 45 dB(A) would be satisfactory in less sensitive rooms such as kitchens, bathrooms and halls. The bedroom is a sensitive room and should not be exposed to more than 35 dB(A). Steps should be therefore be taken to eliminate the potential problem.

Recommendations

Figure 120 shows an alternative ground-floor layout which eliminates the potential noise problems identified above. The bedroom has been moved away from the pump room wall and replaced by the kitchen, which is less sensitive. The resulting layout gives improved stacking as the plan type B now nearly matches that of type A on the floor above.

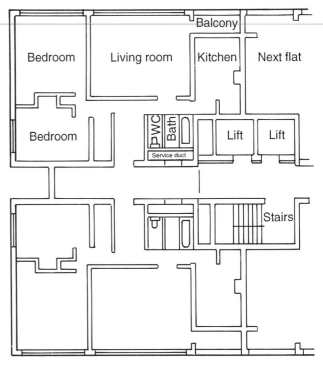

Flat type A
Upper-floor plan

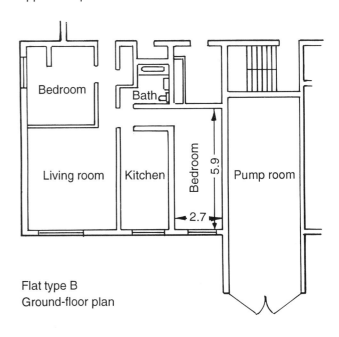

Flat type B
Ground-floor plan

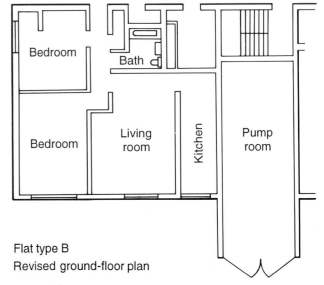

Flat type B
Revised ground-floor plan

Figure 120

Calculation of pump noise in flat type B

The plan of the pump room is shown in Figure 119 and the section is shown on Figure 121. Flanking transmission will be well controlled by virtue of three features of the building design:

- The separating wall extends beyond the face of the building.

- The pump room has its own independent ceiling, which reduces flanking transmission via the common floor slab.

- The rooms sit on a solid concrete ground slab.

It is therefore reasonable to assume that the main noise path is directly through the separating wall.

The appropriate formula, taken from page 13, is:

$$R = D + 10 \log (S/A)$$

where

R = Sound reduction index of wall (dB) (See page 28 for wall data)

D = Level difference, $L_1 - L_2$ (dB)

L_1 = Sound level, pump room (dB)

L_2 = Sound level, bedroom (dB)

S = Shared area of the wall (m²) ($5.9 \times 2.4 = 14.16$ m²)

A = Total area of absorption, bedroom (m²) (10 m² assumed)

$10 \log (14.16/10) = 1.5$ dB

Rearranging the formula:

$$L_2 = L_1 - R + 10 \log (S/A)$$
$$L_2 = L_1 - R + 1.5 \text{ dB}$$

As the octave band spectrum of the source is known, the calculation should be carried out using octave band data. Details are given in Table 17.

The resulting noise level in the adjacent room is 38.4 dB(A).

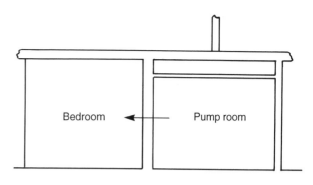

Figure 121

Table 17

	Octave band centre frequency (Hz)					
	125	250	500	1000	2000	'A'
Noise level measured in pump room L_1	91	87	79	69	61	
A-weighting	−16	−9	−3	0	+1	
A-weighted pump noise	75	78	76	69	62	
Total level (L_1 dB(A))*						82
Sound reduction index, 480 kg/m² brick wall (see page 30)	−41	−45	−50	−53	−58	
10 log (S/A)	+1.5	+1.5	+1.5	+1.5	+1.5	
Sound level in bedroom L_2	35.5	34.5	27.5	17.5	5.5	
Total level (L_2 dB(A))*						38.4

* Where the levels are given to greater accuracy than the nearest dB, either round the numbers to the nearest whole dB and use the method on page 7 or use the following formula to calculate the composite sound level:

$$L_{comp} = 10 \times \log (\text{antilog } L_1/10 + \text{antilog } L_2/10 + \text{etc})$$

Design of a quiet room

Brief

A client wishes to have the construction of her planned new house amended to provide one room where she can work in quiet conditions or listen to music without disturbing the rest of the household. The proposed ground- and first-floor house plans are shown in Figure 122.

- Identify the room which is most suitable for this purpose.

- What form of construction should be used for the partitioning, and how can flanking transmission be controlled?

Selection of quiet room

Bedroom 4 is most suitable for conversion to a quiet room for the following reasons:

- It is located over the garage, which is not sensitive to noise and is not a source of frequent noise.

- The bedroom partitions coincide with external masonry walls below, which could support increased mass, if necessary.

- Bedroom 4 is remote from all water services.

Two main building elements must be considered in designing bedroom 4 as a quiet room:

- the partition construction between bedrooms 3 and 4, and

- the partition and door between bedroom 4 and the staircase.

Partition construction to bedroom 3

The partition construction should be capable of providing sound insulation of at least 48 dB $D_{nT,w}$ when installed (see page 27). The formula given on page 13 should be used to express this requirement in terms of the weighted sound reduction index, R_w. Sound-insulation values for masonry partitions are given in Table 15 on page 39 and, for timber-frame partitions, in Table 16 on page 51.

The necessary calculations are given in the box on this page. They indicate that the sound reduction index of the partition should be at least 47.4 dB R_w. The following forms of construction, taken from Tables 15 and 16, meet or very nearly meet this standard:

Construction

	Approximate R_w (dB)
102.5 mm brickwork, plastered both sides (268 kg/m²), field result	47
115 mm autoclaved aerated concrete blockwork, dry-lined (95 kg/m²)	47
70 mm metal channel with two layers of 12.5 mm plasterboard both sides	48
100 mm timber studs with two layers of 12.5 mm plasterboard, one side fixed via resilient channels, 25 mm absorbent quilt in the cavity	50

Flanking transmission

There are three main flanking paths:

- Via the inner leaf of the external wall

- Via the floor (assumed to be timber-joist construction)

- Via the ceiling/roof void

A masonry partition can best control the first two transmission paths. It can be bonded into the inner leaf of the external wall to restrain it, and carried through the floor construction, supporting the floor joists on hangers on both sides of the wall. Control of flanking transmission via the ceiling and roof void is considered in more detail on the next page.

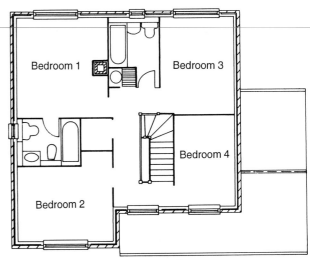

First-floor plan

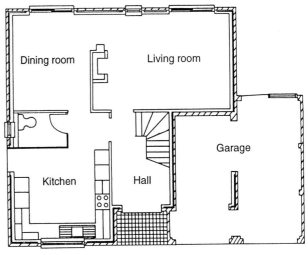

Ground-floor plan

Figure 122

Calculation of R_w requirement from $D_{nT,w}$ standard
(See pages 13 and 16)

Formula:

$$R_w = D_{nT,w} + 10 \log S/0.32V \text{ dB}$$

where

S = surface area of partition = 5.9 m²

V = receiving room volume (m³)
(bedroom 4: 21 m²)
(bedroom 3: 25.9 m²)

Therefore, for transmission from bedroom 3 to bedroom 4:

$$R_w = 48 + 10 \log (5.9/0.32 \times 21)$$

$$R_w = 48 - 0.6 \text{ dB}$$

$$R_w = 47.4 \text{ dB}$$

For transmission in the opposite direction (bedroom 4 to bedroom 3), this procedure gives a result of $R_w = 46.5$ dB

Control of flanking transmission via the ceiling
There are two possible routes for flanking transmission via the ceiling:

● along the plasterboard and ceiling joists, and

● through the ceiling, into the roof space, and back through the neighbouring ceiling.

The arrangement shown in Figure 123 is similar to the situation where a solid masonry separating wall meets an external timber-frame wall. This detail is shown 'in plan' in Figure 55(b) on page 37. The timber frame and lining are butt-jointed to the separating wall, and the junction is sealed to achieve reasonable sound insulation. The lining is at least 12.5 mm plasterboard, and the separating wall would give sound insulation of at least 52 dB $D_{nT,w}$ (4 dB more than recommended for a quiet room). Figure 123 transfers this arrangement to the ceiling. The important details are:

● Plasterboard at least 12.5 mm thick should be used, and the junction between ceiling and partition should be sealed.

● If possible, the ceiling and joists should not be carried across the partition. Where the joists must carry through, the isolation should be improved by fixing the ceiling in the quiet room to the joists via resilient bars (see also page 67).

This design should be checked for flanking transmission, through the ceiling, into the roof space, and back through the neighbouring ceiling. The sound reduction index of a single sheet of 12.5 mm plasterboard is 28 dB R_w (see Table 16 on page 51). Assuming that R_w is approximately equal to $D_{nT,w}$, 28 dB loss occurs as sound travels from bedroom 4 into the roof space, and a further 28 dB loss occurs as sound travels from the roof space into bedroom 3. A total loss of approximately 56 dB would be achieved, and this is 8 dB higher than recommended for quiet rooms. Therefore, flanking transmission via the roof space is not expected to prevent the criterion from being met, as long as there are no apertures in the ceiling, and the edges are well-sealed.

Partition and door to stairs
The sound insulation of a partition between a room and the adjoining hall or corridor will be influenced mainly by the sound insulation of the door. A solid-core timber door with good seals will give a performance of approximately 30 dB R_w (see Table 14 on page 30.) A minimal partition construction of 75 mm timber studs with one layer of 12.5 mm plasterboard on both sides gives a better sound-insulation performance of 36 dB R_w.

Figure 49 on page 31 can be adapted to calculate the composite sound insulation of the door and partition. Assuming that the door area is 20% of the total partition area, the resulting sound insulation is approximately 32 dB R_w (Figure 124).

If the level difference between bedroom 4 and the stairs is 32 dB, a further 16 dB is required to meet the criterion of 48 dB sound insulation between the quiet room (bedroom 4) and the nearest affected bedroom across the hall (bedroom 2). This modest requirement can be achieved by an unsealed hollow-core door (approximately 17 dB R_w. See Table 14 on page 30).

As long as the bedroom doors are closed, the quiet room requirements will be met.

Section

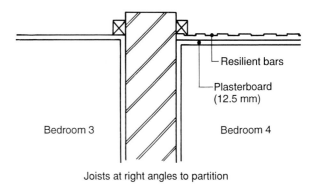

Joists at right angles to partition

Section

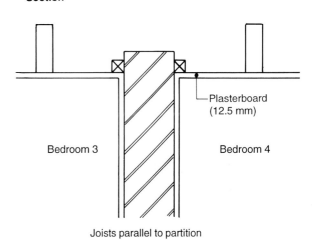

Joists parallel to partition

Figure 123

See Table 14 for typical R values

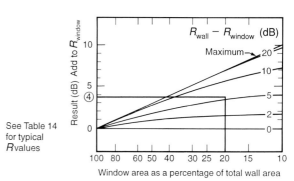

Figure 124

Conversion properties

Conversion of a terraced house

Brief

Figure 125 shows the plans of a two-storey terraced house which is to be subdivided to create two one-bedroom flats.

- Devise plans which will minimise possible future problems of noise from neighbours.

- Identify the sound-insulation requirements between rooms.

- Assess the suitability of the existing construction to meet the sound-insulation requirements.

- Where the existing construction is not adequate, specify alternative, more suitable forms of construction.

- Specify measures to minimise noise disturbance from circulation areas.

Existing building construction

Wall construction

- Separating walls 215 mm brickwork, plastered both sides

- External walls 215 mm brickwork, plastered on the inside

- Brick partitions between rooms 1 and 2 and between WC and kitchen 103 mm brickwork, plastered both sides

- Stud partitions 100 mm timber studs, with wood-lath and plaster finish both sides

Floor/ceiling construction

- Front part of the building (ceiling height 2.8 m) 225 mm × 50 mm timber joists. 22 mm plane-edge floorboards nailed to joists. Wood-lath and plaster ceiling (Figure 125(c))

- Rear extension (ceiling height 2.4 m) 200 mm × 50 mm timber joists, 22 mm plane-edge floorboards nailed to joists, ceiling two layers of 9.5 mm plasterboard (Figure 125(d))

Joists run front-to-back.

(a)
Ground floor before conversion

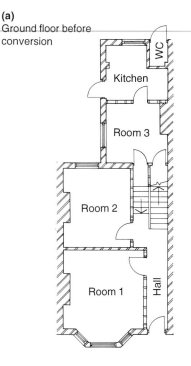

(b)
First floor before conversion

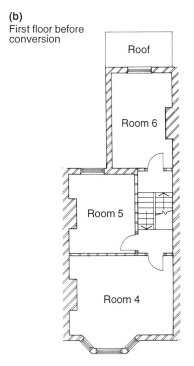

(c)

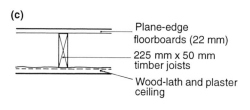

Plane-edge floorboards (22 mm)

225 mm × 50 mm timber joists

Wood-lath and plaster ceiling

(d)

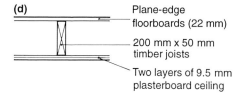

Plane-edge floorboards (22 mm)

200 mm × 50 mm timber joists

Two layers of 9.5 mm plasterboard ceiling

Figure 125

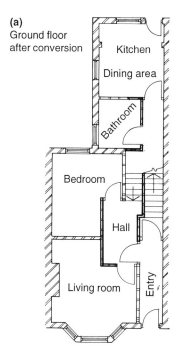

(a)
Ground floor
after conversion

Kitchen

Dining area

Bathroom

Bedroom

Hall

Living room

Entry

(b)
First floor
after conversion

Roof terrace

Kitchen

Bathroom

Bedroom

Living room

Figure 126

Planning to control internal noise

Plans must ensure compatibility of use between adjacent rooms in neighbouring dwellings. Where possible, aim for similar occupancy in neighbouring flats. In particular, flats for large families should not be situated immediately above or below small flats or bedsits.

Handing

Handing of house plans is the usual arrangement in terraced houses of this type. When converting a group of houses, similar floor plans should be retained throughout if the brief permits it. When converting an individual house, the planning should be a handed layout of the neighbouring property, if possible.

Stacking

Identical ground- and first-floor plans are not usually feasible because of the need for circulation space on the ground floor. Figure 126 shows a practicable arrangement providing two one-bedroom flats using the principle of stacking similar rooms on top of each other. The principle breaks down mainly in the following locations:

● The first-floor living room is partly over the ground-floor entrance lobby (see 'Circulation areas' below).

● The first-floor kitchen and new roof terrace are above a more sensitive kitchen/dining room on the ground floor.

Buffer zones

The ground-floor hall provides a buffer zone which protects sensitive rooms in the ground-floor flat from the noise of first-floor tenants using the stairs to their flat. It also provides two doors between the entrance lobby and any sensitive room in the ground-floor flat.

Where appropriate, a fitted wardrobe between a bedroom and living room can also serve as a buffer zone within a flat.

Circulation areas

The entrance lobby shares a common wall with the ground-floor living room and a common floor/ceiling with the first-floor living room. In this instance, there are only two tenancies and the potential for disturbance is small. In larger converted properties, the members of a number of households may use one entrance lobby, giving a disproportionate amount of disturbance to the ground-floor flat.

Services

All water services are contained in the rear extension away from sensitive bedrooms and living rooms. The best location for a water tank is the bathroom if there is space. Water pipes (particularly those carrying mains-pressure water) should be fixed to the external brickwork wall rather than the brick partition wall between bathroom and bedroom or the new bathroom stud partition.

Sound-insulation requirements

England and Wales

Where an existing wall, floor or stair is to become a separating element between dwellings, it is necessary either to show that it already meets the requirements or to adopt a treatment which will bring it up to standard. A construction will meet the requirements if either of the following can be demonstrated:

● It is generally similar to one of the Approved Document new-build constructions (for example within 15% of the mass of a construction in Part C of this manual)

● The construction is shown to meet the field test performance standards given in Table 4 on page 18

There are two ways to meet the requirements if it cannot be demonstrated that the existing construction already meets them:

● Use one of the constructions reproduced in Part C, pages 64 to 71. When using them, there is no need to demonstrate that a given numerical performance will be achieved.

● Repeat an alternative treatment which has been built and tested in a building or a laboratory and which meets the numerical performance standards given on page 18.

The worked examples which follow are treated as if the building were situated in England or Wales.

Northern Ireland

It is anticipated that the Regulations will be extended to cover flat conversions in early 1994. Technical Booklet G1 will contain deemed-to-satisfy provisions for sound insulation where an existing wall or floor becomes a separating wall or a separating floor. The requirements are expected to be broadly as described for England and Wales, though not all the proposed floor treatments will be adopted and no treatment will be given for stairs.

Scotland

Flat conversions in Scotland are subject to the same requirements as new-build dwellings (see page 19). One additional floor construction is offered. It is designed for use in conversion properties.

Partition walls and floors

There are no regulatory requirements. Suggested design standards are given on page 27.

Separating walls

1 Existing separating wall between dwellings

(Figure 127)
Construction: 215 mm brickwork, plastered both sides

This construction is similar to construction A on page 36. As long as the wall mass is not less than 319 kg/m² and the construction is free from flaws, it will satisfy Approved Document E. The wall should be checked for constructional flaws. (Refer to checklists on page 39.)

It is not normally necessary to treat a wall of this type or to carry out sound-insulation testing. Care should be taken in the design and construction of the conversion works not to impair the sound insulation of the wall. (Refer to checklists on page 39.)

2 Separating wall between ground-floor living room and communal lobby

(Figure 127)
Construction: 100 mm timber studs. Wood-lath and plaster finish to both sides

This wall construction is not similar to an Approved Document construction and its field airborne sound insulation performance is likely to fall well short of the Approved Document values:

Approximate partition performance: 42 dB $D_{nT,w}$
(Estimated from laboratory test value on page 51; see box)

Approved Document minimum field value: 49 dB $D_{nT,w}$

The Approved Document offers one wall treatment: an independent leaf with absorbent material (see page 64). The treatment must be applied to both sides of the partition in situations where the existing wall is not of masonry construction at least 100 mm thick with plaster finish on both faces. In this example, there is not enough space for treatment on the entrance hall side of the partition. Consequently, the Approved Document treatment cannot be adopted. Any alternative specification should be cleared with the building control officer. Adoption of an Approved Document new-build construction should prove acceptable. The recommendations which follow use the principles of the timber-frame separating wall shown on page 48.

Recommendations

If the existing stud partition is in good condition, the wood-lath and plaster should be removed from the room side only. An absorbent quilt should be installed in the cavity, new studwork erected and a 30 mm plasterboard lining fixed to it to create a cavity of 200 mm. Details are shown on Figure 128.

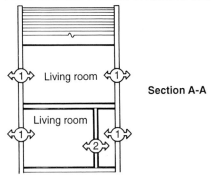

Calculation of $D_{nT,w}$ from laboratory result
(See pages 13 and 16)
Published laboratory sound insulation for partition is 38 dB R_w (see page 51)

Formula

$$R_w = D_{nT,w} + 10 \log S/0.32V \text{ dB}$$

where

S = surface area of wall = 6.4 m²

V = receiving room volume = 47 m³

Therefore

$$38 = D_{nT,w} + 10 \log (6.4/0.32 \times 47)$$

$$38 = D_{nT,w} - 4 \text{ dB}$$

so

$$D_{nT,w} = 38 + 4 = 42 \text{ dB}$$

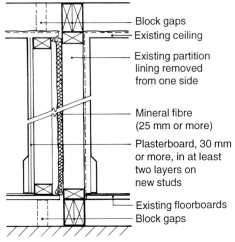

Section A-A

Figure 127

Block gaps
Existing ceiling

Existing partition lining removed from one side

Mineral fibre (25 mm or more)

Plasterboard, 30 mm or more, in at least two layers on new studs

Existing floorboards
Block gaps

Figure 128

3 Separating wall between dwellings (ground-floor flat)

(Figure 129)

Construction: 100 mm timber studs. Wood-lath and plaster finish to both sides

The partition which is to be introduced between the hall and the stairs performs the function of a separating wall between dwellings. The stairs themselves also perform the function of a separating element, and should be designed accordingly.

Recommendations

If practicable, the new wall should be built using a new-build construction from the Approved Document. Solid brickwork plastered both sides or concrete blockwork plastered both sides (construction A or B page 36) would be most appropriate. As this is a conversion property, it may be acceptable to reduce the mass requirements by up to 15%. If this construction is not considered practicable, further relaxations from new-build constructions may be negotiable with the building control officer on the grounds that the wall separates two non-habitable areas within the flats.

Suitable treatments for stairs are given in the Approved Document (see page 70). In this case, it is convenient to provide an under-stair cupboard. The existing stair treads should receive a soft covering, such as carpet, 6 mm thick. (The lease for the flat should provide for the maintenance of this finish.) The underside of the stair should be lined using plasterboard, at least 12.5 mm thick. A small, heavy, well-fitted cupboard door should be fitted (see Figure 130).

4 Separating walls between dwellings (first-floor flat)

(Figure 131)

Construction: 100 mm timber studs. Wood-lath and plaster finish to both sides

This partition supports the upper half of the stairs and separates a noise-sensitive bedroom from a first-floor circulation area. This gives rise to complaints, particularly of impact noise, in many converted properties.

Recommendations

The existing wall is a stud partition. According to the Approved Document, an independent leaf should be installed on both sides of this partition. The resulting reduction in the width of the stairs makes it impracticable in this case. As the common partition area is small, it would not be unreasonable to install an independent lining on the bedroom side only, or to adopt the treatment shown in Figure 128. As the stair treads are in contact with the partition, the soft finish recommended should be taken to the top of the stairs.

5 Hall to ground-floor flat

(Figure 132)

Sound transmission through lobbies should be sufficient to prevent a shortfall in the performance of separating walls and floors. The flat entrance door and doors to sensitive rooms off the hall should have mass of at least 25 kg/m^2. The entrance doors should be well-sealed (see page 73). The inner doors need be hung only to a normal good fit.

6 Entrance hall

(Figure 132)

The banging caused by doors which open onto the entrance hall should be controlled by fitting a closer which reduces the speed of impact, or by fitting compression seals to the frame. A carpet should be specified in the hall to reduce impact sounds at source and to provide some acoustic absorption.

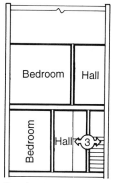

Section B-B

Figure 129

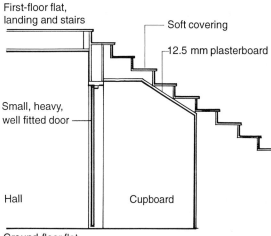

First-floor flat, landing and stairs

Soft covering

12.5 mm plasterboard

Small, heavy, well fitted door

Hall

Cupboard

Ground-floor flat

Figure 130

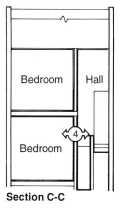

Section C-C

Figure 131

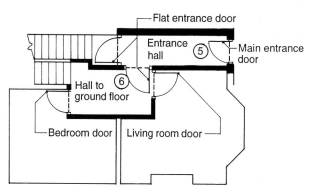

Figure 132

7 Separating floor between dwellings (front part of the building)

(Figure 133)

Construction: 225 mm × 50 mm joists. 22 mm plane-edge floorboards. Wood-lath and plaster ceiling

This is a typical untreated older house floor. With favourable flanking conditions, it might just meet the field test requirements for airborne sound, but it is unlikely to comply with the impact requirements (see page 18):

Floor values (page 68): 46 to 49 dB $D_{nT,w}$ / 67 to 69 dB $L'_{nT,w}$

AD field values: 48 dB $D_{nT,w}$ / 65 dB $L'_{nT,w}$

Recommendations

In order to comply with the Approved Document, floor treatments 1 (independent ceiling with absorbent material), 2 (floating layer, platform floor) or 3 (ribbed floor), reproduced on pages 68 and 69 should be adopted. See Figures 134 and 135. The choice will depend on whether it is more practicable to reduce the height of the ceiling or to raise the level of the floor. Two alternative treatments are available where a strong case can be made for not using floor treatments 1, 2 or 3 (see page 70).

Flanking transmission

The Approved Document does not require detailed consideration of flanking transmission. Guidance on the control of flanking transmission to achieve new-build standards is given on page 61.

8 Separating floor between first-floor living room and communal lobby

(Figure 133)

Construction: As for 7 above

Recommendations

As for 7 above.

9 Separating floor between dwellings (rear part of the building)

(Figure 136)

Construction: 200 mm × 50 mm timber joists. 22 mm plane-edge floorboards nailed to joists. Ceiling two layers of 9.5 mm plasterboard

This untreated floor construction falls well below the Approved Document field test requirements for airborne or impact sound (see page 18):

Typical floor values 38 dB $D_{nT,w}$ / 73 dB $L'_{nT,w}$

AD field values 48 dB $D_{nT,w}$ / 65 dB $L'_{nT,w}$

Recommendations

Although in England and Wales there is no explicit requirement for impact sound insulation from external areas such as terraces, it would be best to adopt a treatment which controls noise from the terrace as well as from the kitchen above. The reduced ceiling height in this part of the building prevents adoption of floor treatment 1 (independent ceiling with absorbent material). Where a strong case can be made for not using floor treatments 1, 2 or 3 (see pages 68 and 69), alternative treatments 4 and 5 are available (see page 70). As the treatment must be carried out from below, treatment type 4, the alternative independent ceiling with absorbent material, is the one to adopt (see Figure 137).

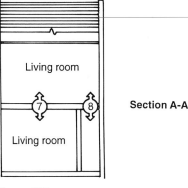

Section A-A

Figure 133

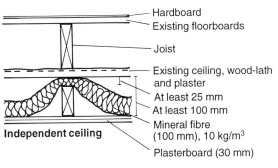

Hardboard
Existing floorboards
Joist
Existing ceiling, wood-lath and plaster
At least 25 mm
At least 100 mm
Mineral fibre (100 mm), 10 kg/m³
Plasterboard (30 mm)

Independent ceiling

Figure 134

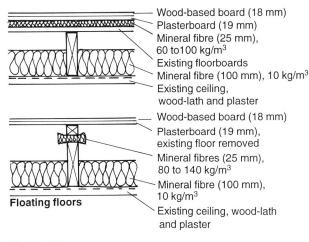

Wood-based board (18 mm)
Plasterboard (19 mm)
Mineral fibre (25 mm), 60 to 100 kg/m³
Existing floorboards
Mineral fibre (100 mm), 10 kg/m³
Existing ceiling, wood-lath and plaster

Wood-based board (18 mm)
Plasterboard (19 mm), existing floor removed
Mineral fibres (25 mm), 80 to 140 kg/m³
Mineral fibre (100 mm), 10 kg/m³
Existing ceiling, wood-lath and plaster

Floating floors

Figure 135

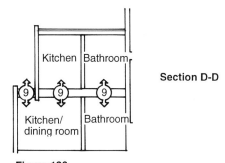

Section D-D

Figure 136

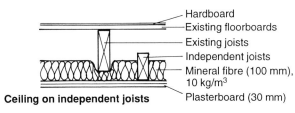

Hardboard
Existing floorboards
Existing joists
Independent joists
Mineral fibre (100 mm), 10 kg/m³
Plasterboard (30 mm)

Ceiling on independent joists

Figure 137

10 Existing stud partition

(Figure 138)

Construction: 100 mm timber studs with wood-lath and plaster finish both sides

This partition construction meets the minimum standard recommended on page 27.

Recommended minimum standard   38 dB $D_{nT,w}$

Approximate partition performance   38 dB $D_{nT,w}$

(Laboratory test, page 51: 38 dB R_w. See page 13 and box on page 90 for conversion between R and D).

11 Existing brick partition

(Figure 139)

Construction: 113 mm brickwork, plastered both sides.

This partition meets the minimum standard recommended on page 27.

Recommended minimum standard: 38 dB $D_{nT,w}$

Approximate partition performance: 46 dB $D_{nT,w}$

(Field test, page 39: 47 dB R'_w. See page 13 and box on page 90 for conversion between R and D).

12 New stud partition between rooms

Figure 139

The required sound reduction index, R_w, must be determined to allow selection of a suitable construction.

Recommended minimum standard   38 dB $D_{nT,w}$

Required R_w 37 dB R_w

(The method of calculation is given in the box.)

Construction: (selected from Table 16, page 51)
75 mm timber studs with one layer of 12.5 mm plasterboard both sides and with 25 mm mineral-fibre quilt in the cavity. Associated sound insulation 40 dB R_w

If it is necessary to resist moisture, the plasterboard layer on the bathroom side may be replaced by dense plaster on expanded metal lath without a reduction in sound insulation (see Figure 140).

13 New stud partition between a room and hall or corridor

(Figure 139)

The sound insulation of a partition between a room and the adjoining hall or corridor will be influenced mainly by the sound insulation of the door. A solid-core timber door installed to a normal good fit will give a performance of approximately 21 dB R_w (see page 30). A minimal partition construction of 75 mm timber studs with one layer of 12.5 mm plasterboard on both sides gives a much better sound-insulation performance of 36 dB R_w (see Table 16 on page 51). Consequently, the partition construction need not be selected on the basis of sound-insulation requirements.

Calculation of R_w requirement from $D_{nT,w}$ standard
(See pages 13 and 16)

Formula

$$R_w = D_{nT,w} + 10 \log S/0.32V \text{ dB}$$

where

S = surface area of partition = 4.1 m², and
V = receiving room volume (kitchen/dining room) (18.7 m³)

Therefore

$$R_w = 38 + 10 \log (4.1/0.32 \times 18.7)$$

$$R_w = 38 - 1 \text{ dB}$$

$$R_w = 37 \text{ dB}$$

First-floor plan

Figure 138

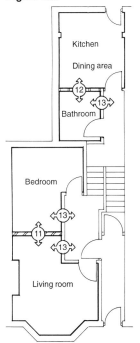

Ground-floor plan

Figure 139

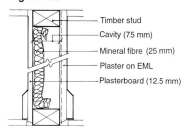

Timber stud
Cavity (75 mm)
Mineral fibre (25 mm)
Plaster on EML
Plasterboard (12.5 mm)

Figure 140

Loft conversion

Brief

A client owns three pairs of cottages and wishes to convert the lofts into bedrooms. Each pair of cottages has a different separating wall construction in the loft space below a slated roof:

Pair A: 215 mm brickwork, terminated below the pitched slated roof

Pair B: 102.5 mm brickwork, terminated below the pitched slated roof

Pair C: No separating wall in the loft space

Show the detailing necessary in each case to ensure reasonable separating-wall sound insulation. How can the sound insulation from the outside be maximised?

Pair A

(See Figure 141)
The basic construction complies with the Approved Document (see also the example on page 90.)

Direct transmission

Make good all the mortar joints and apply a plaster finish or dry lining to the existing wall to bring it into line with construction A, page 36.

Flanking transmission

Adopt the principles given for the control of flanking transmission around solid masonry walls (see page 37). Fill the junction between the top of the wall and the roof, and fix plasterboard to the rafters.

Roof sound insulation

The following methods should be adopted to improve the sound insulation against external noise:

● Ensure that there is an imperforate membrane beneath the roofing slates.

● Include an absorbent quilt in the cavity to improve acoustic isolation between the two skins.

● Increase the mass of the ceiling layer by fixing additional layers of plasterboard.

● For better results, improve the mechanical isolation between roof and ceiling layers by one of the following methods:

1 Fix the plasterboard ceiling to the rafters via resilient bars (see page 67)

2 Install independent timber studs between existing rafters, and fix the plasterboard ceiling to these only

Pair B

(See Figure 142)
To comply with the Approved Document, an independent leaf with absorbent material should be applied to one side (see page 64). For better results, apply the treatment to both sides, as shown in this example.

Direct transmission

Make good all the mortar joints, and install a panel on both sides comprising 25 mm plasterboard with staggered joints, supported top and bottom only and creating a cavity at least 25 mm wide. Install an 25 mm absorbent quilt of at least 10 kg/m³ in each cavity.

Flanking transmission
As for pair A.

Roof sound insulation
As for pair A.

Pair C

(See Figure 143).
As there is presently no separating wall, this should be treated as a new-build wall.

Direct transmission

Either continue the existing 215 mm brickwork separating wall through (see pair A), or adopt a timber-frame separating wall construction (see page 48).

Flanking transmission
As for pair A.

Roof sound insulation
As for pair A.

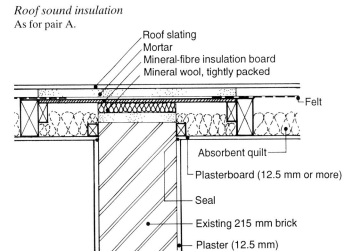

Roof slating
Mortar
Mineral-fibre insulation board
Mineral wool, tightly packed
Felt
Absorbent quilt
Plasterboard (12.5 mm or more)
Seal
Existing 215 mm brick
Plaster (12.5 mm)

Figure 141

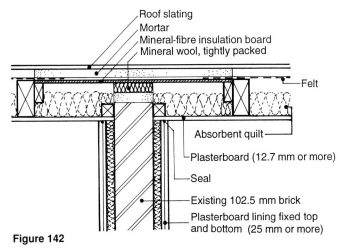

Roof slating
Mortar
Mineral-fibre insulation board
Mineral wool, tightly packed
Felt
Absorbent quilt
Plasterboard (12.7 mm or more)
Seal
Existing 102.5 mm brick
Plasterboard lining fixed top and bottom (25 mm or more)

Figure 142

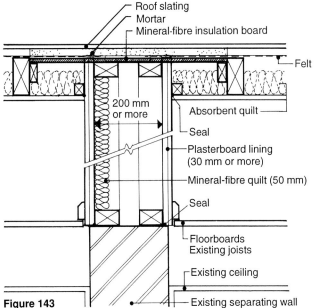

Roof slating
Mortar
Mineral-fibre insulation board
Felt
200 mm or more
Absorbent quilt
Seal
Plasterboard lining (30 mm or more)
Mineral-fibre quilt (50 mm)
Seal
Floorboards
Existing joists
Existing ceiling
Existing separating wall

Figure 143

Noise problems in existing dwellings

Fourteen self-contained examples deal with noise problems which arise in existing dwellings. They fall into five main groups:

- *Outside-to-inside sound insulation*
  Road traffic, aircraft and other external sources

- *Separating-element sound insulation*
  Masonry and lightweight separating walls, concrete and timber-joist separating floors

- *Specially insulated rooms*
  Quiet room and music room

- *Noise nuisance and the law*
  Legal remedies to noise problems

- *Domestic mechanical noise sources*
  Domestic machine, water pump and lift

Each example gives a typical brief followed by a technical solution to the problem, including details of any calculations necessary to solve it.

Urban flat subject to road traffic noise

Brief
A flat owner complains of road-traffic noise in his bedroom which overlooks a busy urban street (see Figure 144). The room is ventilated by opening the window, which increases the disturbance. Make practical recommendations to maximise the outside-to-inside sound insulation while maintaining adequate ventilation. Estimate the reduction in traffic noise which can be achieved, compared with the present situation with windows closed.

Site inspection
The wall construction is 215 mm solid brickwork, plastered on the inside. The window occupies 50% of the wall area. It is a single-glazed vertical sliding sash. The window reveal is 150 mm wide (see Figure 145). There are no gaps between frame and surround, and inspection of the wall indicates that there are no penetrations through the wall construction.

Recommendations
The remedial measures fall into two parts:

● Maximise the sound insulation of the window.

● Provide alternative means of ventilation.

The window
The sound insulation of the existing window can be increased by installing secondary windows in the existing opening. The following actions will maximise the final result:

● Enclose as large an air gap as possible. Allowing for the frame, an air gap of about 100 mm should be practicable in this case.

● Use reasonably thick glass in the secondary window, preferably not less than 6 mm thick.

● Ensure that both casements make an airtight seal with their frames. Fit rubber or neoprene wipe seals to the outer window. Select an inner window with casements which have or can accept rubber or neoprene compression seals. Bed the new frame in mastic to ensure a good seal with its surround.

● If practicable, fix an acoustically absorbent material to at least three of the window reveals. The higher its absorption coefficient the better, preferably at least 0.7 at 500 Hz and above.

Ventilation
A ventilator should be installed (or a pair of ventilators) to meet the requirements of the Noise Insulation Regulations, 1975, for airflow, self noise and sound insulation against external noise (see page 33).

Noise reduction
The simple computation procedures and data given on pages 30 and 31 have been used to estimate the sound insulation before and after treatment (see box). The sound insulation should increase by approximately 7 dB(A) from 29 to 36 dB(A). In practice, such schemes have typically given 34 dB(A) sound insulation and up to 40 dB(A) in the best cases.

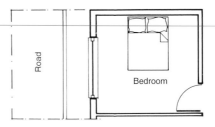

Figure 144

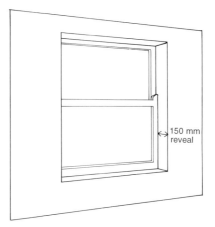

Figure 145

Estimate of sound insulation before and after treatment
(See pages 30 and 31)

Figure 146

	$R_{A(traffic)}$ (dB(A))
Before treatment	
Wall sound insulation	47
6 mm single glazing (good fit)	26
R (wall) − R (window)	21
50% glazed area, add to R (window). See graph	3
Composite sound insulation	29
After treatment	
Wall containing ventilator	39
Double window (6, 100, 6 mm)	34
R (wall) − R (window)	5
50% glazed area, add to R (window). See graph	2
Composite sound insulation	36

Estimated increase in sound insulation:
$$36 - 29 = 7 \text{ dB(A)}$$

House subject to road traffic noise
Brief
A two-storey house is situated 22 m from the kerbside of a busy road (see Figure 147). The intervening land is grassed. The owner would like to erect a barrier no more than 3 m high to reduce noise levels at the house.

● Where on the site is the most effective barrier location?

● What reduction in traffic noise can be expected at a ground-floor window 1.2 m above the ground and at a first-floor window 4.5 m above the ground?

● What are the important constructional considerations for barrier attenuation?

Solution
Barrier location
The closer the barrier to the road, the more effective it will be. Consequently, the barrier should be situated at the site boundary closest to the road and should extend along its entire length. The line of sight between the house and the road should be blocked completely . If the road can still be seen beyond the ends of the site frontage, it will be necessary to return the barrier along the sides of the site (see Figure 147)

Traffic noise reduction
The reduction in road-traffic noise should be computed using *Calculation of road traffic noise* (see page 10). There are two stages of calculation in this case:

● Removing the effect of the grassed land. The additional attenuation associated with grassed land is lost when a barrier is introduced.

● Assessing the barrier attenuation.

For the ground-floor window:

	(dB(A))
Grassland attenuation lost	
Distance from edge of carriageway 22 m	
Receiver height above ground 1.2 m	
Soft-ground correction (see Figure 148)	4

Barrier attenuation
Path difference = a + b – c (See Figure 149)

$$a = \sqrt{5.5^2 + 2.5^2} = 6.04 \text{ m}$$

$$b = \sqrt{20^2 + 1.8^2} = 20.08 \text{ m}$$

$$c = \sqrt{25.5^2 + 0.7^2} = 25.51 \text{ m}$$

a + b – c = 0.61 m

Barrier attenuation (see Figure 150)	14
Resulting overall reduction = 14 – 4 =	10

For the first-floor window:

Grassland attenuation lost	1
Barrier attenuation	12
Resulting overall reduction = 12 – 1 =	11

Barrier construction
The barrier material should be imperforate. To ensure that sound transmitted through the barrier will not prevent the barrier achieving the required attenuation of 14 dB(A), the surface mass of the barrier material should be at least 10 kg/m².

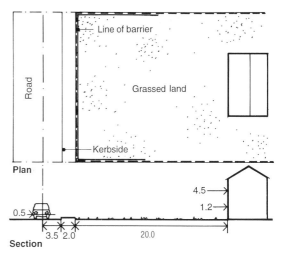

Figure 147

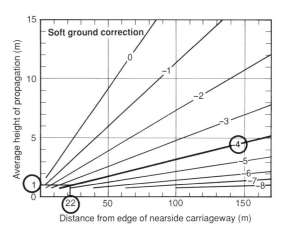

Figure 148

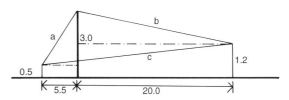

Path difference = a + b – c (m)

Figure 149

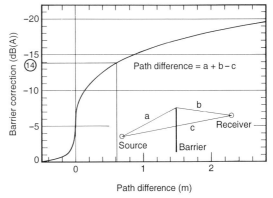

Path difference (m)

Figure 150

House subject to aircraft noise

Brief

The owner of a house under the flight path to a busy airport complains of aircraft noise in his first-floor bedroom. The room is ventilated by opening the window, which increases the disturbance. Make recommendations to improve the outside-to-inside sound insulation of the building envelope by approximately 10 dB(A) while maintaining adequate ventilation.

Site inspection

The wall construction is 215 mm solid brickwork, plastered on the inside. The window occupies 40% of the external wall area. It is a 4 mm single-glazed window installed to a good airtight fit. The internal window reveal is 150 mm. The roof is pitched with tiles on roofing felt on rafters. The ceiling is a single sheet of 9 mm plasterboard on joists (see Figure 151).

Assessment method

The simple approach proposed for road-traffic noise on page 30 will give results which are typically within about 3 dB of more accurate methods. The formula and its interpretation when used in aircraft noise calculations are given in the box on this page.

Recommendations

The remedial measures fall into two parts:

● Improve the sound insulation of the window and the roof.

● Provide an alternative means of ventilation.

Window

The sound insulation of the window can be increased by installing double windows in the existing opening. Examples of calculations are given on page 99, using data from Table 14 on page 30. This shows that two 6 mm panes separated by a 100 mm cavity (34 dB $R_{A(traffic)}$) will give the desired result. The constructional principles listed on page 31 should be adopted to maximise the final result in practice.

Roof

The calculation procedure indicates that a roof with an $R_{A(traffic)}$ value of approximately 38 dB will provide adequate sound insulation. This corresponds to a pitched roof with a wood-lath and plaster ceiling with a 100 mm absorbent quilt. To obtain a similar result with modern materials, a ceiling with similar mass should be adopted, for example two layers of plasterboard, giving a total thickness of at least 32 mm, plus skim coat.

Ventilation

The chosen method of ventilation should not impair the sound insulation of the building envelope. One available method is to install a ventilator (or pair of ventilators) which meets the requirements of the Noise Insulation Regulations, 1975, for airflow, self noise and sound insulation against external noise (see page 32). The device should be fitted in accordance with the manufacturer's recommendations.

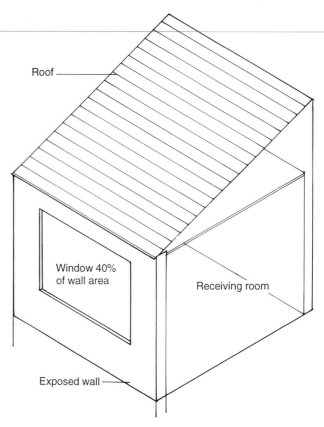

Figure 151

Formula

$$L_2 = L_1 - R_{A(traffic)} \ (dB(A))$$

where:

L_2 is the sound-pressure level received in the room

L_1 is the aircraft noise level on site (but away from reflecting objects)

$R_{A(traffic)}$ is the sound reduction index of the building element expressed in dB(A) and based on a typical road-traffic noise spectrum

Method

1 Obtain separate $R_{A(traffic)}$ values for the facade and roof from Table 14, page 30. In this case, the facade value is the composite insulation of the wall and window together. Assess this using Figure 49, page 31. If more than one facade is exposed, consider each separately.

2 Subtract each $R_{A(traffic)}$ value from the external sound level in turn. If you do not know the external sound level (L_1), choose an arbitrary value for ease of calculation. This is valid only for estimating the building envelope sound insulation. If you want to estimate the actual level in the improved receiving room, measure the sound level in dB(A) on site and substitute the value in the formula.

3 Combine the calculated levels using the rules for addition of decibels given on page 7 to give the total level received. Subtract this from L_1 to give the overall building envelope sound insulation (which is not influenced by your choice of L_1).

4 Improve the value of the weakest element in the building envelope and recalculate the overall sound insulation.

5 Repeat step 4 until the result reaches the desired value. The example on page 99 gives the results of the first and last calculations in this process.

Example	
Before improvement	**dB(A)**
External sound level (arbitrary)	100
Wall sound insulation (215 mm plastered brick)	47
Window sound insulation (4 mm single glazing)	24
R(wall) – R(window)	23
Window area 40% wall area.	
Figure 152 below gives result	4
Add to window value for composite wall value	28
Sound level received via wall (100 – 28)	72
Roof sound insulation	
(Pitched tiled with 9 mm plasterboard ceiling)	27
Sound level received via roof (100 – 27)	73
Combined sound level	
(72 and 73; difference 1, add 3 to higher)	76
Combined sound insulation (100 – 76)	24

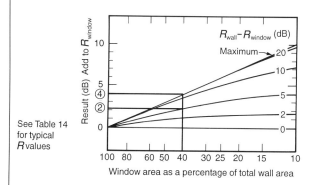

See Table 14 for typical R values

Figure 152

After improvement	dB(A)
External sound level (as above)	100
Wall sound insulation (as above + ventilator)	39
Window sound insulation (6, 100, 6 mm)	34
R(wall) – R(window)	5
Window area 40% wall area.	
Figure 152 above gives result	2
Add to window value for composite wall value	36
Sound level received via wall (100 – 38)	64
Roof sound insulation	
(As above but with quilt on 32 mm ceiling)	38
Sound level received via roof (100 – 38)	62
Combined sound level	
(64 and 62; difference 2, add 2 to higher)	66
Combined sound insulation (100 – 66)	34
Overall improvement (34 – 24)	10

Houses subject to other noise sources

The simple approach to outside-to-inside sound insulation calculations is valid for road-traffic noise because the $R_{A(traffic)}$ values for building envelope elements are based on a typical road-traffic noise frequency spectrum. The previous example shows how the method can be adapted to estimate aircraft sound levels. Sound-insulation improvement methods already given for road traffic and aircraft noise will reduce the level of other external noise sources, such as railways and factories, by a similar degree. The exact result will, however, depend on the frequency content of the source. The designer must have this information before he or she can make accurate predictions of the sound insulation resulting from changes to the building envelope. The principles of this more involved design method are given in the box here and illustrated in the worked examples on pages 115 and 117.

More detailed approach to sound-insulation prediction
Formula

$$R = L_1 - L_2 + 10 \log S - 10 \log A + 6 \text{ dB}$$
(Derived from formula on page 13, Sound reduction index (loudspeaker))

L_1 is the source sound level (dB)
Measure typical source levels 1 m outside the exposed facade(s). Make a calibrated tape recording and analyse it in octave bands. Obtain either the L_{eq} over a representative time period, or typical maximum levels. The frequency range 125 Hz to 2000 Hz will be adequate for many sources, but extend this range if the source sound contains much energy at lower or higher frequencies. Remove the effect of the sound reflection from the building facade by subtracting 3 dB at each frequency.

L_2 is the received sound level (dB)
Measure the corresponding sound level averaged in the receiving room. A-weight the octave band values and sum using the rules for addition of dB on page 7. The result is the received dB(A) level. Compare this with an appropriate internal noise criterion to assess the reduction required (see page 20).

S is the surface area of facade element(s) (m²)

A is the area of absorption in the receiving room (m²)
Measure the reverberation time, *T*, in the receiving room. (See pages 12 and 13 for definition. *T* is measured by recording a pistol shot in the room and measuring the slope of the reverberant decay.) If the receiving room is a typical furnished living room or bedroom, assume that *T* = 0.5 seconds at all frequencies. Measure the room volume, *V* (m³), and compute *A* using the formula:

$$A = \frac{0.16 \times V}{T} \quad (\text{m}^2)$$

Sound insulation of masonry separating wall

Brief

A flat owner complains of noise from the neighbouring flat. He complains that he can hear his neighbours' TV set and hi-fi and that he can hear voices, even during normal conversations. He provides an architect's drawing from which constructional details are taken (see Figure 153). Carry out a preliminary appraisal of the problem. Advise on the likely causes and on what action should be taken to remedy the problem.

Preliminary appraisal

Nature of the disturbing noise

TV, radio and voices are sources of airborne sound. The client can hear neighbours' voices at normal conversational levels, suggesting that the airborne sound insulation may be deficient.

Appraisal of drawings

A simple analysis (see 'Constructional appraisal' on this page) demonstrates that the separating wall shown on the drawing has insufficient MASS and ISOLATION to provide sound insulation in line with new-build constructions in England and Wales, Northern Ireland and Scotland (see pages 18 and 19). The flanking constructions comply with detailing given in supporting documents to the various Regulations and need not be improved for the construction as a whole to comply with new-build standards.

Site inspection

A site inspection should be conducted along the following lines:

● Examine the building construction to confirm that the separating wall and flanking constructions were built according to the design drawings. (Follow the site inspection checklist on page 39.)

● Strip off some of the separating wall lining to discover whether the brick wall was built without gaps in the mortar jointing. (Sound-insulation tests may be necessary if there is no clear evidence to support the claim that the sound insulation is deficient or if the information is required for legal purposes. Advise the client to have field tests carried out according to the latest edition of British Standard BS 2750 before the existing wall construction is disturbed during site inspection.)

● Check the junction between the top of the brickwork and the floor slab. It is particularly difficult to seal this junction.

Practical constraints

The client places the following constraints on the remedial works, and they rule out the adoption of an Approved Document new-build construction.

● The work must be carried out in the client's property only.

● The reduction in width of the affected room must be minimal.

● The increase in weight of the separating wall must be kept to a minimum.

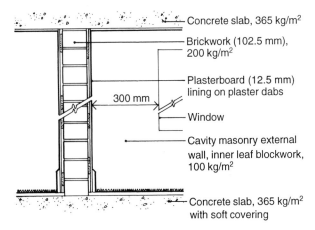

Concrete slab, 365 kg/m²
Brickwork (102.5 mm), 200 kg/m²
Plasterboard (12.5 mm) lining on plaster dabs
300 mm
Window
Cavity masonry external wall, inner leaf blockwork, 100 kg/m²
Concrete slab, 365 kg/m² with soft covering

Figure 153

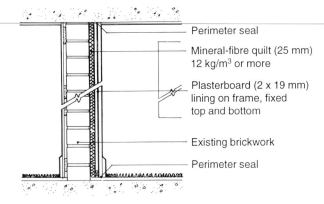

Perimeter seal
Mineral-fibre quilt (25 mm) 12 kg/m³ or more
Plasterboard (2 x 19 mm) lining on frame, fixed top and bottom
Existing brickwork
Perimeter seal

Figure 154

Constructional appraisal

Mass

This type of separating wall is covered on pages 36 to 39 and is similar to construction B from Approved Document E. The main difference is the reduced MASS of the masonry core in this example, which corresponds approximately to a 4 dB reduction in sound insulation, averaged over the frequency range 100 Hz to 3150 Hz (see Figure 17, page 14).

Critical frequency

The use of reduced mass in the masonry core also causes its CRITICAL FREQUENCY to rise to around 200 Hz, where the resulting dip in performance is more damaging (see page 14).

Isolation

The lightweight plasterboard linings must be well ISOLATED from the masonry core if they are to contribute significantly to its airborne sound insulation. In this case, the mechanical isolation is poor, because of the dabs which connect the elements rigidly. The acoustic isolation is poor because there is no absorbent material in the gap and, more seriously, because the choice of panel mass and airgap causes a MASS-AIR RESONANCE at 173 Hz (see box for calculation). This is close to the critical frequency of the masonry core and is likely to result in a large dip in performance in the 125 Hz octave band.

Mass-air resonance

$$f_{res} = \frac{1900}{\sqrt{m \times d}}$$

where:

m = panel mass = 10 kg/m²

d = cavity width = 12 mm

Mass-air resonance, f_{res} = 173 Hz

Flanking transmission

The flanking wall constructions comply with the detailing recommended for new-build separating walls (see page 37).

Recommendations

An independent leaf and absorbent material should be adopted (see page 64).

● Strip off the dry lining from one side of the separating wall.

● Repoint any poor jointing in the brickwork paying particular attention to the junction between the wall and the slab above. Seal with mastic if movement cracking is likely.

● Install a self-supporting laminated plasterboard partition 25 mm from the separating wall, enclosing a 25 mm mineral-fibre quilt (see Figure 154)

Sound insulation of lightweight separating wall
Brief
The owner of a flat in a converted warehouse complains of noise disturbance from the neighbouring flat. He can hear his neighbours' voices, even at normal conversational levels. The following information is available:

● Architect's drawings of the flats (Figure 155)

● The results of sound-insulation tests carried out by the environmental health officer (Figure 156)

● The results of laboratory sound-insulation tests carried out by the wall-system manufacturer (Figure 157).

Decide what a reasonable standard of sound insulation would be, and compare this with the field and laboratory measurements. Advise on the likely causes of the poor performance, and specify measures to remedy the problem.

Criteria
The various Building Regulations give an indication of what represents 'reasonable' sound insulation. The minimum acceptable mean sound-insulation value through separating walls between new dwellings is approximately 52 dB $D_{nT,w}$. In England and Wales a lower standard of 49 dB $D_{nT,w}$ is accepted in flat conversions. Northern Ireland is expected to adopt provisions similar to those of England and Wales in 1994. New-build standards apply to conversions in Scotland (see pages 18 and 19).

Field test results
The standardised level difference (D_{nT}) has been measured in one-third octave bands between 100 Hz and 3150 Hz. In Figure 156, the weighted standardised level difference is obtained using the the procedure given on page 16.

Result: 46 dB $D_{nT,w}$.

Laboratory results
In Figure 157, the weighted sound reduction index (R_w) is obtained from the laboratory results using the procedure given on page 16. The result must be converted to a $D_{nT,w}$ value using the formula given on page 13 (see box):

Result: 57 dB $D_{nT,w}$.

Effect of flanking transmission
Flanking transmission is likely to prevent a value of 57 dB $D_{nT,w}$ from being reached on site. However, the flanking elements are not responsible for the poor field performance because, in general, a shortfall due to flanking transmission would tend to affect the result at all frequencies. In this case, the failure is in the mid-to-high part of the frequency range.

Cause of the failure
Gaps through partitions cause a reduction in sound insulation at high frequencies. The pronounced dip measured in this case is due to the situation shown in Figure 158. There is a gap under the wall, but it is partly covered by skirtings, creating two small voids in which sound resonates. The problem can be remedied by removing the skirtings and filling the voids behind with sealant to make them airtight before replacing the skirtings.

Calculation of $D_{nT,w}$ from laboratory result
Formula

$$R_w = D_{nT,w} + 10 \log (S/0.32V)$$

where S = Surface area of wall (7.2 m²)
            V = receiving room volume (36 m³)

Therefore $55 = D_{nT,w} + 10 \log (7.2/0.32 \times 36)$
             $55 = D_{nT,w} - 2$

So $D_{nT,w} = 55 + 2 = 57$ dB $D_{nT,w}$.

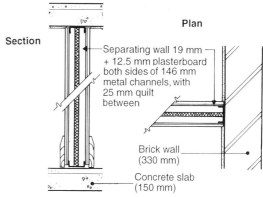

Figure 155

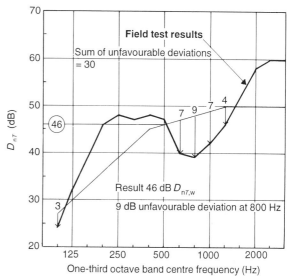

Figure 156

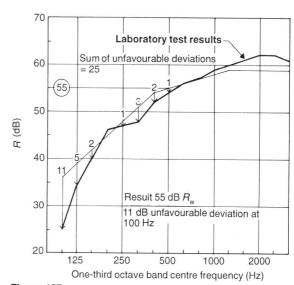

Figure 157

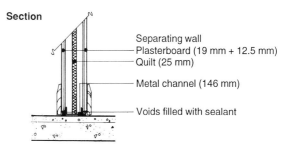

Figure 158

101

Sound insulation of concrete separating floor

Brief

Residents in a four-storey block of flats collectively complain of poor sound insulation. Their main complaints are as follows:

● Each resident is disturbed by TV, hi-fi and voices from the flats immediately above and below.

● There is also some disturbance from the flat two floors above or two floors below (see Figure 159).

Carry out a preliminary appraisal of the problem. Advise on the likely cause and the remedial steps which should be taken.

Preliminary appraisal

It is unlikely that the residents would be in such agreement if the problem were simply one of noisy neighbours. This suggests that the airborne sound insulation is generally deficient. Reports that sounds from remote floors can be heard suggest that flanking transmission is involved.

Appraisal of drawings

Figure 160 shows a typical section through a separating floor and the external wall. The floor construction complies with one of the specifications recommended in Approved Document E and equivalent regulatory documents for Northern Ireland and Scotland (see page 56). However, the junction between the floor and the external wall fails to comply with the Approved Document on two counts:

● The concrete floor should pass through the inner leaf of the external wall.

● The first joint in the concrete floor should be at least 300 mm away from the wall cavity.

Site inspection

A site inspection should be conducted to ensure that the separating floor has been built according to the design drawings. Some of the floating floor should be taken up to inspect the junction with the external wall to ensure that the space between the final beam and the external wall has been filled with concrete.

Recommendations

The following remedial actions are recommended:

Separating floor

● Take up the floating floor as necessary to make good any cracking or gaps in the concrete floor.

● Replace the floating floor to the original specification. (Edge details are shown in Figure 161.)

Flanking wall
(Figure 161)

● It will not be practicable to comply with the detailing given for concrete floors in the Approved Document. Use the method given for timber floors with a lightweight solid masonry flanking wall (see pages 44 and 61).

● Strip off the existing dry lining.

● Fix a 25 mm absorbent blanket to the wall.

● Install a plasterboard lining, fixed on a frame at top and bottom only; either:

    1 plasterboard/cellular core, 18 kg/m², joints taped between panels and at edges, or

    2 two sheets 12.5 mm plasterboard with staggered joints.

● Leave a gap of at least 3 mm between skirting and floating floor. Seal with acrylic caulk or neoprene.

● Turn the resilient layer up at the edge to isolate the floating layer from the wall.

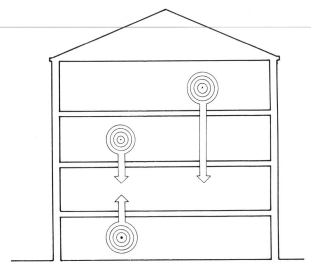

Figure 159

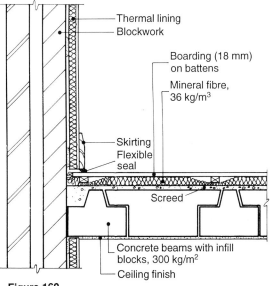

Thermal lining
Blockwork

Boarding (18 mm) on battens

Mineral fibre, 36 kg/m³

Skirting
Flexible seal

Screed

Concrete beams with infill blocks, 300 kg/m²

Ceiling finish

Figure 160

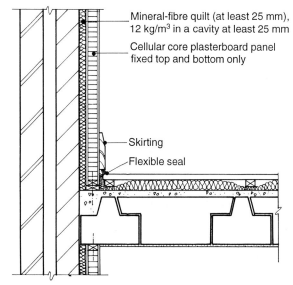

Mineral-fibre quilt (at least 25 mm), 12 kg/m³ in a cavity at least 25 mm

Cellular core plasterboard panel fixed top and bottom only

Skirting

Flexible seal

Figure 161

Sound insulation of timber separating floor
Brief
The owner of a converted flat complains of noise from the flat above. He is disturbed mainly by footsteps and furniture being dragged across the floor. He can also hear his neighbour's hi-fi but cannot hear voices during normal conversations. He sends a drawing showing the floor construction (Figure 162). The external walls are built of 215 mm brickwork and the partitions are of timber studs lined with plasterboard. There is no carpet in the living room of the upstairs flat. The environmental health officer carries out an impact sound test in accordance with British Standard BS 2750 Part 7. The results are shown in Figure 163.

Assess the impact sound insulation. Does it achieve a reasonable standard? Is it in line with the type of construction used?

Field test results
The standardised impact sound-pressure level (L'_{nT}) has been measured in one-third octave bands between 100 Hz and 3150 Hz. The procedure given on page 17 should be used to obtain the weighted standardised impact sound-pressure level $L'_{nT,w}$ (see Figure 163).

Result: 65 dB $L'_{nT,w}$.

Criteria
The various Building Regulations (see page 18) give an indication of what represents 'reasonable' sound insulation. The maximum acceptable weighted standardised impact sound-pressure level for separating floors in new dwellings is approximately 61 dB $L'_{nT,w}$. In England and Wales and in Northern Ireland, a higher level of 65 dB $L'_{nT,w}$ is accepted in flat conversions.

Therefore, the floor in question just meets the standard associated with flat conversions.

Appraisal of drawings
Figure 162 shows a typical section through the separating floor. The floor and associated flanking constructions comply with one of the specifications given in Approved Document E (see construction B, page 56). Therefore, this construction should be capable of meeting new-build standards, rather than conversion standards.

There are no faults in the general design of the floor/ceiling construction which would account for the reduced impact sound insulation. The shortfall must be attributed either to poor workmanship or to subsequent deterioration.

Site inspection
A site inspection should be conducted and the following checks made:

- Ensure that the separating floor has been built according to the design drawings.

- Look for bridging of the floating floor by nails, trunking, skirtings, etc.

- Take up some of the floating floor to examine the resilient quilt. Check that it is continuous under the battens, of the correct density, and has not become wet.

Cause of failure
The shortfall in sound insulation in this case was caused by the use of nails to secure the floating floor to the joists. A total of 12 nails had been used. Bridging of this type could have occurred during construction if supervision were inadequate. DIY activity by the resident is another possible cause of unwanted bridging across floating layers.

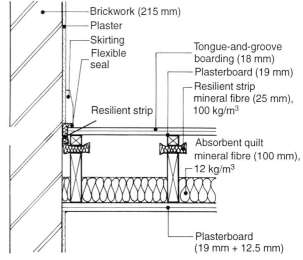

Figure 162

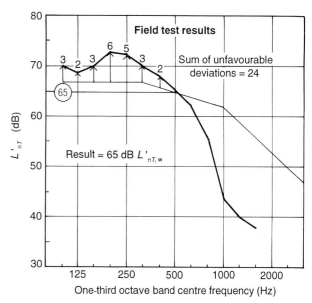

Figure 163

Quiet room

Brief

A house owner wishes to use one of his first-floor bedrooms as a study, and to make it reasonably free from disturbance from other parts of the dwelling (see Figure 164). He wants any building work to be restricted to the study only.

- How much sound insulation does he need?

- How much sound insulation does the existing building construction provide?

- How can the sound insulation be improved to meet a good standard?

Existing construction

The partitions are generally 75 mm timber studs with one layer of 12.5 mm plasterboard on each side. The bathroom/study partition construction and the floor construction are shown in Figure 165.

Criteria

On page 27, the following minimum sound-insulation standards are recommended for quiet rooms:

Partition walls: 48 dB $D_{nT,w}$
Partition floors: 46 dB $D_{nT,w}$

Existing building construction

Partition to bathroom

The partition performance will lie between that of construction numbers 3 and 5 in Table 16, page 51, around 40 dB R_w.
Surface area = 0.32 × volume. Therefore,

$$D_{nT,w} = R_w.$$

Estimated performance: 40 dB $D_{nT,w}$.

Partition to bedroom

The sound insulation depends on two elements:

Partition 36 dB R_w (number 3 Table 16, page 51)

Cupboard door in bedroom 2: 17 dB R_w (Table 14, page 30).

The combined effect will easily meet the quiet room partition criterion above.

Partition floor to kitchen

Estimated performance: 38 dB $D_{nT,w}$ (see page 68).

Flanking transmission

115 mm autoclaved aerated concrete blockwork, dry- lined; performance: 47 dB R_w (Table 15, page 39)

Flanking transmission via this wall is not expected to prevent the design targets from being achieved.

Study door

Hollow-core door: 17 dB R_w (Table 14, page 30)

Recommendations

The following remedial actions are recommended:

Partition to bathroom

- Remove the lining from one side of the partition.

- Install plasterboard on resilient channels (page 67) to improve mechanical and acoustical isolation and to increase mass (see Figure 166).

Estimated performance: 50 dB $D_{nT,w}$, ignoring flanking transmission.

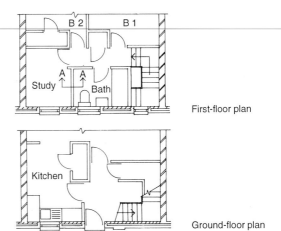

Figure 164

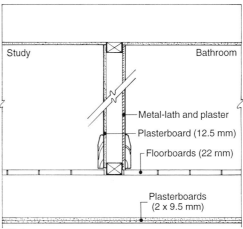

Kitchen
Section A-A

Figure 165

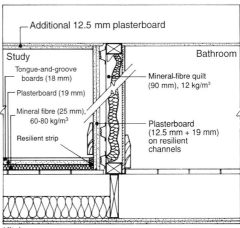

Kitchen
Section A-A
Figure 166

Partition floor to kitchen

- Adopt construction A, page 60, which has improved mechanical and acoustical isolation and increased mass (see Figure 166). As the existing ceiling has lower mass than construction A, its performance will be somewhat lower than shown on page 60. Allow 3 dB (based on Figure 17, page 14)

Estimated performance: 50 dB $D_{nT,w}$, ignoring flanking transmission.

Study door

- Install a solid-core timber door (25 kg/m^2) with compression seals all round (see page 73)

Estimated performance: 30 dB R_w.

Music room

Brief

The house owner in the previous example wants his study to double as a music practice room for his daughter, who plays the trombone. How can he protect his next door neighbour?

Criteria

An estimate should be made of the additional sound insulation required from the separating wall. Do this by finding out how loudly the girl plays the trombone, and how this compares with normal domestic activities. Figure 167 is a graph showing typical maximum sound-pressure levels measured from a trombone in a small room. The lower curve is a typical maximum sound-pressure level from a domestic TV set or radio. Each spectrum should be A-weighted and the values combined, using the rules for addition of decibels on page 7 (see box).

Result

Trombone	109 dB(A)
TV/Radio	70 dB(A)
Difference	39 dB(A)

(This procedure exaggerates the dB(A) values, as all frequencies are not produced simultaneously.)

Separating walls should normally give at least 52 dB $D_{nT,w}$ sound insulation. Therefore, to reduce the level of the trombone to that of a domestic TV or radio next door, a level difference of 91 dB $D_{nT,w}$ (52 + 39) will be required.

Walls giving this degree of sound insulation are massive, consume a large amount of space, employ specialised methods of construction and are seldom built, even in broadcasting and recording studios. The method given on page 64 could improve the sound insulation of separating walls by perhaps 10 to 15 dB, but only if flanking transmission can be controlled. Clearly, there is no complete technical solution to a problem of this kind.

In practice, the client's daughter is unlikely to play the trombone loudly all the time. With 10 to 15 dB improvement in sound insulation, and some restriction on the hours spent practising, the potential disturbance can be kept to a minimum.

Recommendations

The sound insulation between properties should be improved by installing an independent lining next to the separating wall and the external flanking wall. The maximum practicable improvement will be obtained if the following principles are followed (Figure 168):

Mass Existing brickwork wall, plastered on both sides. Independent lining, mass as high as possible, consistent with structural and practical considerations

Isolation No physical contact between the lining and the existing wall. The larger the cavity the better. Absorber, 25 mm mineral fibre, density at least 10 kg/m³.

See Figure 169 for a possible solution. The window should be closed during practice, to control flanking transmission via the open air.

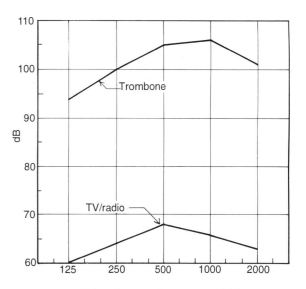

Octave band centre frequency (Hz)

Figure 167

Calculation of dB(A) values					
Frequency (Hz)	125	250	500	1000	2000
Trombone (dB)	94	100	105	106	101
A-weightings	−16	−9	−3	0	+1
A-weighted trombone	78	91	103	106	102
78 and 91 (difference 13, add 0)			91		
91 and 102 (difference 11, add 0)			102		
102 and 103 (difference 1, add 3)			106		
106 and 106 (difference 0, add 3)			109		
Answer	109 dB(A)				

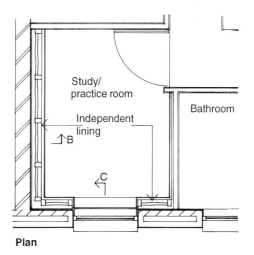

Plan

Figure 168

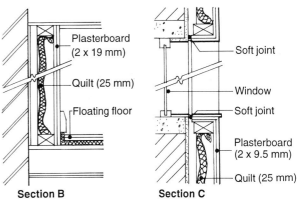

Section B   Section C

Figure 169

Noise nuisance and the law
Brief
A house owner complains of noise from the pub next door (see Figure 170). He can hear the jukebox playing all evening, and is awakened every night by people slamming car doors in the car park. What steps can he take to eliminate the disturbance?

Approach to the problem
Architectural remedies
If the complainant wishes to take action on his own property, he can adopt any of the measures described previously to control external noise:

● A noise barrier along the site boundary (see page 97)

● Secondary glazing and mechanical ventilation in the affected rooms (see pages 96, 98 and 99).

Legal remedies
Whenever the character or level of a disturbing noise causes unreasonable interference in the home, there are legal remedies available. These are outlined below.

Noise nuisance
Legal provisions fall into two categories.

Common law
The complainant may bring a private noise nuisance action in the civil court, usually the county court in the first instance. If the case is proven, the possible remedies include damages and an injunction against the person responsible.

Statutory law
Noise is a statutory nuisance under the provisions of the Environmental Protection Act, 1990. Section 80 describes the steps which local authorities are to take where they are satisfied that a noise nuisance exists. The procedure is broadly as follows:

● The complainant informs the local environmental health department that he or she is suffering a noise nuisance.

● An environmental health officer (EHO) visits the premises to make an assessment.

● If the assessment confirms that the noise is a nuisance, the officer is empowered to serve an abatement notice on the person responsible. The notice will normally require the person responsible to take specified steps to abate the nuisance within a given period.

● The person served with the notice has 21 days in which to appeal to the magistrate's court. The defence of 'best practicable means' is available to the noise maker if operating from industrial, trade or business premises.

● If the person served with the notice contravenes any of its requirements without reasonable excuse, then that person is guilty of an offence and liable to a fine in the magistrate's court. If this does not provide an adequate remedy, the local authority can take action in the High Court to secure abatement, prohibition or restriction of the nuisance. Alternatively, if the person responsible defaults, the local authority has powers to carry out any works required by the notice.

There may be circumstances preventing the EHO from taking action. For example, the EHO may not consider the noise to be a nuisance, or the local authority itself may be responsible for the noise. Section 82 of the Act enables the complainant to bring proceedings in the magistrate's court. If the court is satisfied that a nuisance exists, it will serve an order against the person responsible requiring him or her to abate the nuisance.

This brief outline of the available legal procedures is illustrated in flow-diagram form in Figure 171.

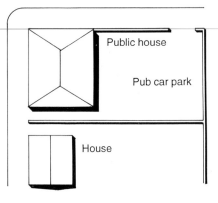

Figure 170

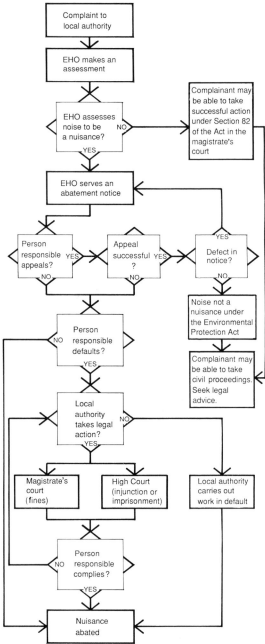

Figure 171

Domestic machine noise

Brief

The owner of a semi-detached house complains of noise caused by his neighbour who works on a sewing machine for many hours each day. The intermittent sound of the machine's motor can be heard equally loudly in the ground-floor living room and in two first-floor bedrooms (see Figure 172). How can the complainant reduce or eliminate the noise in his house?

Criteria

The noise in question is not caused by normal domestic activity. Therefore, it is not valid to consider the problem in terms of 'reasonable' or 'adequate' sound insulation as required by the Building Regulations in England and Wales, Northern Ireland and Scotland.

The problem is to reduce the machine noise to an acceptable level. The numerical value of this acceptable level depends on the individual circumstances. Often, the complainant will not be satisfied until the noise is inaudible, in which case the designer should aim to reduce the disturbing noise to 10 dB below background, if practicable.

Approach to the problem

Whenever the character or level of a disturbing noise causes unreasonable interference, there are legal remedies available which are summarised in the previous example. This example provides a technical remedy.

If the neighbour is willing to co-operate, it is much better to reduce the noise at source. Otherwise, remedial works will be necessary in three rooms in the client's home. Before embarking on detailed investigations, ensure that the machine is not unnecessarily noisy as a result of lack of maintenance. Find out whether a quieter model is available. This could simply and inexpensively resolve the problem to the satisfaction of both parties.

The best method to control noise from the existing machine depends on the way the machine noise is transmitted into the separating wall. There are two possible mechanisms (see page 32):

- Airborne sound transmission, where the sound of the machine is transmitted through the air in the source room to the separating wall. An independent lining next to the separating wall in the source room would reduce airborne sound transmission.

- Structureborne sound transmission, where the machine vibration is transmitted along the floor into the separating wall. This is often the dominant transmission mechanism for machine noise. Clearly, an independent wall lining in the source room would have no effect on this type of transmission.

To find out which mechanism dominates, place the machine temporarily on a thick, soft, resilient material such as a foam rubber cushion. Run the machine, and listen next door. If the level is reduced, then structureborne transmission is at least partly to blame and the machine should be isolated from the surrounding structure.

Figure 173 shows the principles of an inertia block which isolates the machine from its surroundings without causing it to vibrate excessively. The higher the mass of the inertia block the better. Normally, its mass should be significantly greater than that of the vibrating machine. Values of up to six times the machine mass are commonly specified. Guidance should normally be obtained from an expert.

A similar arrangement could be used for a washing machine, drier or other floor-mounted domestic appliance. (Further guidance on washing machines is given on page 34).

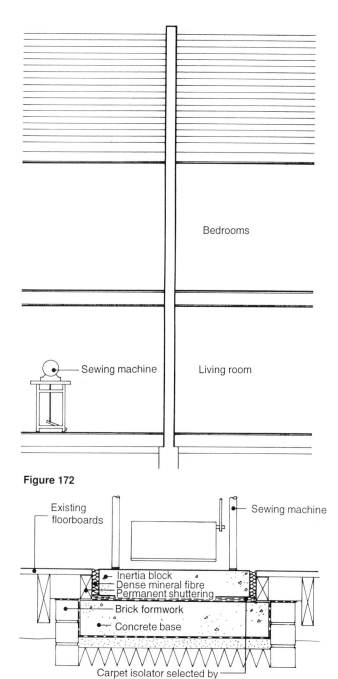

Figure 172

Figure 173

Pump noise

Brief

A home owner installs a water pump on a shelf in his airing cupboard to improve domestic water pressure (see Figure 174). The pump proves noisy in operation, so he replaces the lightweight door on the airing cupboard by a heavy door with good seals. This makes no difference to the noise heard around the house. What steps can he now take to reduce noise from the existing pump?

Sound-transmission mechanism

By fitting a heavy door with good seals onto the airing cupboard, the client has improved the airborne sound insulation of the airing cupboard (assuming that there are no holes or gaps in construction elsewhere). This has not reduced the pump noise, so structureborne sound transmission must be the problem. This occurs when a vibrating system is in direct contact with the surrounding building construction (see page 32).

The vibrating pump is in contact with the building construction in the following places:

- The pump is sitting directly on the shelf which is supported on the stud partition.

- The water pipes to and from the pump are rigidly attached to the stud partition by the pipe brackets.

- Vibration can be transmitted also via any rigid electrical conduits connected to the pump.

Recommendations

The noise of the pump can be reduced by mechanically isolating it from the building construction, and by ensuring that items are not attached to lightweight building elements. Figure 175 shows an ideal arrangement. The following features have been incorporated:

- The pump has been removed from the lightweight shelf and relocated above the heavier concrete floor slab.

- The pump has been turned through 180° so that the pipes can be attached to the brickwork wall rather than the lightweight partition.

- The pump has been fixed to a concrete inertia block to increase the mass of the vibrating system for stability.

- The whole assembly is seated on rubber isolators. A specialist supplier will usually select these on the basis of the mass of the assembly and the rotational frequency of the pump.

- To prevent vibration transmission via the pipes, flexible connectors have been inserted immediately before the first pipe support. These will be effective for pressures up to about 1 bar.

For smaller pumps, such as those used in domestic central heating systems, inertia blocks are not usually necessary, but the manufacturer may be able to provide suitable vibration isolators.

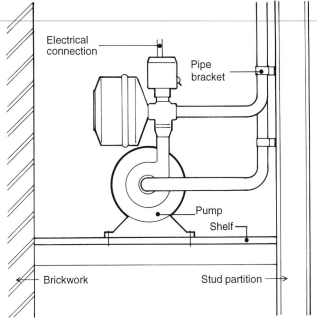

Figure 174

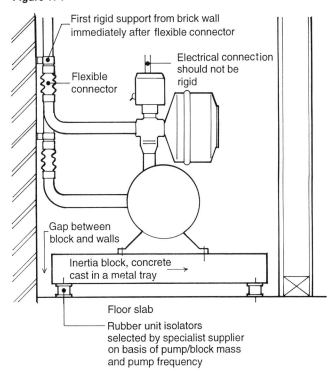

Figure 175

Lift noise
Brief
A flat owner is disturbed by lift noise. Whenever the lift is used, he can clearly distinguish each part of its operation. When the lift is called, he hears a short burst of noise followed by continuous motor noise as the lift moves up or down the shaft. When the lift stops, there is another short burst of noise followed by clicking sounds. When the lift stops on his floor, he can hear the doors opening and closing.

The flat is on the top floor of the building and shares a dense concrete block wall with the lift motor room (see Figure 176). The following information is available:

- Results of noise measurements in the lift motor room and the complainant's flat. Figure 177 shows the octave band levels recorded during the continuous motor noise. Bursts of noise give up to 10 dB increase at each frequency.

- Results of laboratory sound-insulation tests for a dense blockwork wall (see Figure 178).

Make a preliminary appraisal of the problem on the basis of this information.

Preliminary appraisal
The effective sound-level difference through the common wall can be obtained by subtracting the sound levels measured in the flat from those measured in the lift motor room. Compare the result with the laboratory measurement for the wall type (see Figure 178). The level difference measured on site is much lower than the laboratory result. Assuming that there are no gross errors in the construction of the wall on site, this indicates that the sound of the lift motor is being transmitted via the building structure.

Site inspection
A visit should be made to site to inspect the isolation of the lift motor. It should comply with the principles given on pages 108 and 107.

Sources of lift noise
The burst of sound noted by the client is associated with the application and release of the brake on the lift motor. Correct isolation of the lift will reduce this noise too.

The client also noted two other sounds:

- A clicking sound, likely to be caused by the lift switchgear, usually housed in a separate cabinet. This can be controlled by mechanically isolating the cabinet and fitting seals to the cabinet door.

- The noise of the lift doors. Regular maintenance can minimise this noise. Doors with a slow operation speed and nylon runners tend to be less noisy.

Remedial measures
Remedial measures, in particular isolation of the lift motor assembly, should be left to the lift manufacturer to implement as the safe operation of the lift may be affected by the proposed noise-control measures.

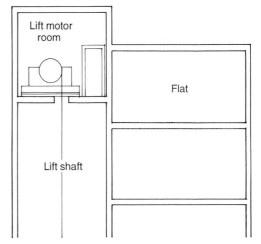

Figure 176

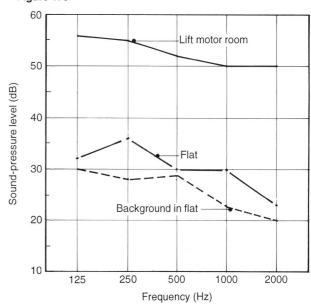

Figure 177

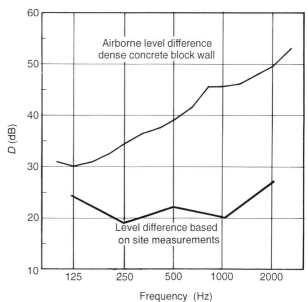

Figure 178

Appendices

Appendix A
Site noise calculations

The more detailed worked examples presented in Appendix A relate to the site for housing, Figure A1, which is the same as that described on page 78.

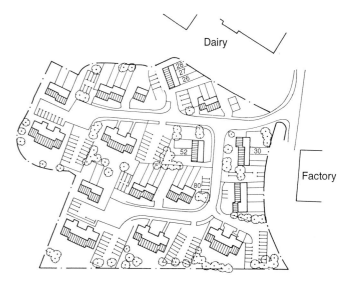

Site plan and plot numbers

Figure A1

Road-traffic noise calculations

Brief 1

Using *Calculation of road traffic noise,* HMSO, 1988 (see page 10), plot the position of the 68 dB(A) L_{10} road-traffic noise contour on site. Calculate the noise levels as if for a first-floor window, 4.5 m above the ground.

Data required

The Highways Department of the local authority provides the following predicted traffic flows on the nearby trunk road for 15 years hence:

● Traffic flow rate 35 000 vehicles per 18-hour day

● Mean traffic speed 74 km/h

● Heavy vehicles 20%

● Road gradient 0°

● Source height (standard) 0.5 m

● Height of receiver point above source 4.0 m

● Average height of propagation 2.5 m (0.5 + 4/2)

The road surface is impervious to surface water. The road is on the same level as the site. The site is flat and grassed. The view of the road from the site is bounded by tall buildings 600 m apart on adjoining sites (see Figure A2).

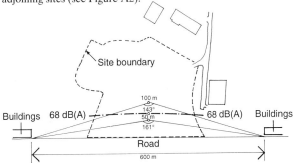

Figure A2

Solution

Try a point 50 m from the edge of the carriageway, about halfway across the site. Its angle of view of the road is 161° (see Figure A2).

	dB(A)
Noise level, traffic flow (Figure A3)	74.4
Correction for mean traffic speed and heavy vehicles (Figure A4)	+3.6
Impervious road surface (for traffic speeds below 75 km/h)	−1.0
Basic noise level	77.0
Distance correction, hard ground (Figure A5)	−6.0
Soft-ground correction, average height of propagation 2.5 m (Figure A6)	−3.1
Angle of view correction (Figure A7)	−0.5
Reflection effect of building facade	+2.5
Predicted noise level	69.9

The 68 dB(A) position must be further from the road. Repeating the calculation for a point 100 m from the edge of the carriageway gives a result of 65.1 dB(A).

The 68 dB(A) position is therefore between 50 m and 100 m. Interpolate between the two distances:

Level change between 50 m and 100 m	4.8
Level change between 50 m and 68 dB(A) point	1.9

Try a distance of 50 + 50 × 1.9/4.8 = 70 m
The recalculated result for a distance of 70 m is 67.7 dB(A).

After further interpolation, the 68 dB(A) L_{10} point is found to be 67 m from the edge of the carriageway. The contour is shown on Figure A2.

112

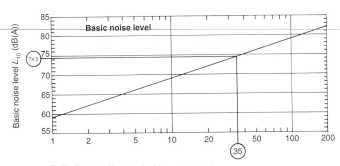

Traffic flow rate (thousand vehicles/18 h day)

Figure A3

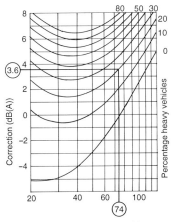

Mean traffic speed (km/h)

Figure A4

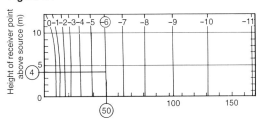

Distance from edge of nearside carriageway (m)

Figure A5

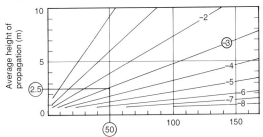

Distance from edge of nearside carriageway (m)

Figure A6

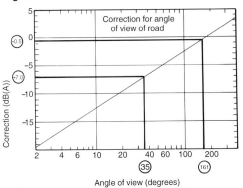

Angle of view (degrees)

Figure A7

Brief 2

Assess the road-traffic noise conditions for plot 80 in terms of the criteria set by the local authority (see pages 78 and 79). How much sound insulation is required from the building envelope? Assess the sound level in the private garden.

Data required

● Road traffic data: as on previous page:

● Source height (standard): 0.5 m

● Height of receiver point above source: 1.5 m

● Average height of propagation: 1.25 m (0.5 + 1.5/2)

Plot 80 is 55 m from the edge of the carriageway. Its view of the road is restricted by high-rise buildings. See Figure A8. Compute the sound level 1 m from the ground and first-floor windows.

Solution: Road-traffic noise on the dwelling	L_{10} (dB(A))
For ground-floor window (height 2 m) basic	
noise level (see previous example)	77.0
Distance correction, 55 m (Figure A5)	–6.4
Soft-ground correction, average height of	
propagation 1.25 m (Figure A6)	–5.1
Angle of view correction 35° (Figure A7)	–7.1
Reflection effect of building facade	+2.5
Predicted facade noise level, ground floor	60.9

For the first-floor window, 4.5 m high, the average height of propagation increases to 2.75 m and the soft-ground correction changes from –5.1 to –3.1 dB; a change of 2.0 dB.
Predicted facade noise level, first floor   62.9

The internal criteria set by the local authority are expressed in terms of dB(A), $L_{Aeq,16h}$. For road-traffic noise (see page 10):

$$L_{Aeq,16h} = L_{A10,18h} - 2 \; (\pm 2 \; dB)$$

Therefore, facade noise levels are approximately:

Ground floor	59 dB $L_{Aeq,16h}$
First floor	61 dB $L_{Aeq,16h}$
Criterion	45 dB $L_{Aeq,16h}$ inside any dwelling

The sound insulation of the building envelope must provide at least 14 dB (59 – 45) for the ground floor and 16 dB (61 – 45) for the first floor. These requirements can easily be met by closed 4 mm single glazing which reduces traffic noise by approximately 24 dB(A). See Table 14 on page 30.

Solution: Road-traffic noise in the garden

Calculate the noise level in the garden, at a point 10 m behind the house (see Figure A9).

	L_{10} (dB(A))
Basic noise level (see previous example)	77.0
Distance correction, hard ground (Figure A10)	
75 m from carriageway, 1.5 m above ground	–7.6
(The additional attenuation caused by grassed	
land is lost when a barrier is introduced.)	
Angle of view correction 35° (Figure A7)	–7.1

Barrier attenuation (see Figure A9)

$$a = \sqrt{63.5^2 + 7.5^2} \quad = \quad 63.94 \text{ m}$$

$$b = \sqrt{15^2 + 6.5^2} \quad = \quad 16.35 \text{ m}$$

$$c = \sqrt{78.5^2 + 1.0^2} \quad = \quad 78.51 \text{ m}$$

Path difference = a + b – c	= 1.78 m	
Barrier attenuation (see Figure A11)		–17.6
Predicted noise level, garden		44.7

Converting to $L_{Aeq,T}$ (see above):

Predicted noise level, garden	43 dB(A), $L_{Aeq,18h}$
Criterion:	65 dB(A), $L_{Aeq,0.5h}$

Therefore, the noise level in the garden is well below the local authority criterion. (In borderline cases, check the result using hourly traffic data.)

Note: Strictly, noise transmission around the sides of the buildings should also be considered. *Calculation of road traffic noise* gives further details.

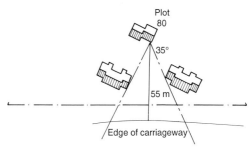

Figure A8

Figure A9

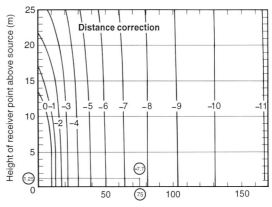

Figure A10

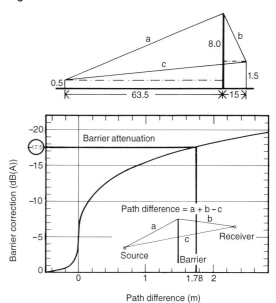

Figure A11

Railway noise calculations

Brief 3

Use the results of sound exposure level, L_{AE}, noise measurements made on site close to the railway track, to compute the following $L_{Aeq,T}$ values:

Day $L_{Aeq,16h}$ (7 am to 11 pm)
Night $L_{Aeq,8h}$ (11 pm to 7 am)
Busiest half-hour $L_{Aeq,0.5h}$

Data required

Five types of train use the nearby track:

(a) Multiple-unit passenger train
(b) Locomotive-hauled passenger train
(c) Locomotive-hauled Speedlink mixed freight train
(d) Double-headed heavy oil train
(e) Light engine

Noise measurements have been made 12 m from the nearest track, just inside the site boundary. The sound exposure level, L_{AE}, has been obtained for each train type (see page 11). The results are recorded in Table A1, along with the information on train movements received from British Rail.

Table A1

Train type	L_{AE} (dB)	Number of train movements (7 am to 11 pm)	Number of train movements (11 pm to 7 am)	Busiest half-hour
(a)	79	125	2	5
(b)	93	65	5	3
(c)	84	35	12	2
(d)	94	5	3	1
(e)	82	10	2	1

Calculation (See page 11)

$$L_{Aeq,T} = 10 \log \left(\frac{\text{antilog } L_{AE1}/10 + \text{antilog } L_{AE2}/10 + ...}{\text{Total time period in seconds}} \right)$$

Daytime assessment (7 am to 11 pm). Consider train type (a) only:

$$L_{Aeq,16h} = 10 \log \left(\frac{\text{antilog } 7.9 + \text{antilog } 7.9 + ...}{57\ 600} \right)$$

There are 125 equal L_{AE} values to consider. Another way to write this is:

$$L_{Aeq,16h} = 10 \log \left(\frac{125 \times (\text{antilog } 7.9)}{57\ 600} \right) = 52 \text{ dB}$$

Repeat this procedure for each train type. The resulting $L_{Aeq,16h}$ values are:

(a) 52, (b) 64, (c) 52, (d) 53, (e) 44

The overall result is obtained by combining these values using the rules for addition of dB given on page 7:

44 and 52 = 53
52 and 53 = 56
53 and 56 = 58
64 and 58 = 65

Therefore, the daytime railway noise level at the reference position is 65 dB $L_{Aeq,16h}$. Repeating this procedure for the night movements gives 59 dB $L_{Aeq,8h}$ and, for the busiest half-hour, 67 dB $L_{Aeq,0.5h}$

The local authority criterion for gardens is 65 dB $L_{Aeq,0.5h}$ during the busiest half-hour. This is 2 dB lower than the level determined at 12 m. Values of $a = 2$ and $r = 12$ can be substituted in the distance formula in the upper box at the foot of this column. The result ($d = 15.8$) indicates that the garden criterion for train noise will be met at locations beyond approximately 16 m from the track.

114

Brief 4

Use the $L_{Aeq,T}$ values obtained in the previous example to assess railway noise conditions for plot 80, in terms of the internal criteria set by the local authority (see pages 78 and 79). How much sound insulation is required from the building envelope?

Data required
(See Figure A12)

Distance between plot 80 facade and track (source–receiver distance)	48 m
Distance between measurement point and track (reference distance)	12 m
Angle of view of track	35°

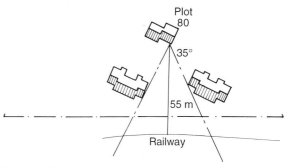

Figure A12

Calculation

	dB(A)
Daytime $L_{Aeq,16h}$, 12 m from track	65
Attenuation is 5 dB per doubling of distance from the track*	
2×12 m = 24 m gives –5 dB	
2×24 m = 48 m gives –5 dB	
Distance attenuation	–10
Angle of view 35°† (Figure A7, page 112)	–7
Facade reflection effect	+2.5
Daytime $L_{Aeq,16h}$ at facade 48 m from track	50.5
Similarly,	
Night-time $L_{Aeq,8h}$ at facade 48 m from track	44.5
Criteria, inside any dwelling	
Daytime $L_{Aeq,16h}$ (7 am to 11 pm)	45
Night-time $L_{Aeq,8h}$ (11 pm to 7 am)	35
Building envelope attenuation needed	
Daytime (50.5 – 45)	5.5
Night-time (44.5 – 35)	9.5

These requirements can easily be met by closed 4 mm single glazing which reduces traffic noise by approximately 24 dB(A), assuming a typical road-traffic noise spectrum (see Table 14, page 30). For borderline situations, assess the facade sound-insulation requirements in octave bands, using an appropriate spectrum (see next example.)

---

* The mathematical formula for this relationship is:
  Distance attenuation, $a = 16.7 \times \log (d/r)$ (dB)
 where
  d = Source–receiver distance (m)
  r = Reference distance (m)
 Alternatively, the distance, d, at which a given attenuation will be obtained, can be calculated using:
  $d = r \times$ antilog $(a/16.7)$ (m)

---

† The following formula should give the same result:
  Angle attenuation = $10 \times \log (\theta/180)$ (dB)
 where
  θ = angle of view (degrees)

Brief 5

Calculate the sound level inside the dwelling on plot 80, assuming that the window must be partly open for ventilation.

Data required

The single-figure methods of calculation given so far will not be accurate enough for these calculations, because they rely on the source being typical road traffic. In this case, the data are required in octave bands. A typical spectrum has been measured during the site noise survey.

Formula

Railway noise behaves much like a line source. Use the equation given on page 30 for road-traffic noise, which is also a line source:

Level difference
$$L_1 - L_2 = R - 10 \log S/A \text{ (dB)}$$

Therefore,
$$L_2 = L_1 - R + 10 \log S/A \text{ (dB)}$$

where

$L_1 =$ Sound level 2 m outside the facade (dB)
(See Table A2 for measured night-time values)

$L_2 =$ Received sound-level in the room (dB)

$R =$ Sound reduction index (dB)
(See Figures 45 and 46 on pages 30 and 31)

$S =$ Surface area, room facade element (m²)
(Assume window = 1.5 m², opening area = 0.5 m² and wall area = 5.5 m²)

$A =$ Absorption in the room (m²)
(This will depend on room finishes. Assume 10 m² at all frequencies, which is a reasonable assumption for a furnished room)

Result of the calculation

Details of the calculation are shown opposite. The result shows that even with the window partly open for ventilation, the internal level does not exceed the local authority criterion of 35 dB(A).

Brief 6

Calculate the sound level in the garden behind plot 80 during the busiest half-hour.

Data required

A typical source spectrum has been obtained during the site noise survey (see Table A2 opposite). Use Figure 26, page 22 to obtain the barrier attenuation in octave bands (see also Figure A13). Assume that the distance correction is now 3 dB per doubling of distance. This is equivalent to hard-ground attenuation (from the *Calculation of road traffic noise*) which must be adopted for barrier calculations (see page 113).

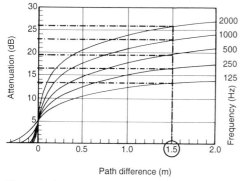

Figure A13

Result of the calculation

Details of the calculation are shown in Table A3 opposite. The result shows that the sound level in the garden does not exceed the local authority criterion of 65 dB(A) during any half-hour.

Table A2

	Octave band centre frequency (Hz)					
	125	250	500	1000	2000	'A'
Calculate the level at plot 80						
Measured noise level, night, 12 m from track	58	43	58	53	53	
A-weighting (See page 7)	−16	−9	−3	0	+1	
A-weighted level	42	34	55	53	54	59
Distance correction (See previous example)	−10	−10	−10	−10	−10	
Angle of view (See previous example)	−7	−7	−7	−7	−7	
Facade reflection (See previous example)	+2.5	+2.5	+2.5	+2.5	+2.5	
Level 2 m from facade L_1	27.5	19.5	40.5	38.5	39.5	44.5
Calculate the reduction in level through each element						
R wall, brickwork	−41	−45	−50	−54	−59	
$10 \times \log (S/A)$ (S = 5.5 m², A = 10 m²)	−2.6	−2.6	−2.6	−2.6	−2.6	
Sound level received through wall only	−16.1	−28.1	−12.1	−18.1	−22.1	
R, window	−20	−23	−26	−29	−29	
$10 \times \log (S/A)$ (S = 1.5 m², A = 10 m²)	−8.2	−8.2	−8.2	−8.2	−8.2	
Sound level received through window only	−0.7	−11.7	6.3	1.3	2.3	
R, open window (assumed)	0	0	0	0	0	
$10 \times \log (S/A)$ (S = 0.5 m², A = 10 m²)	−13.0	−13.0	−13.0	−13.0	−13.0	
Sound level received through opening only	14.5	6.5	27.5	25.5	26.5	
Combine the levels*						
Via wall	−16.1	−28.1	−12.1	−18.1	−22.1	
Via window	−0.7	−11.7	6.3	1.3	2.3	
Via opening	14.5	6.5	27.5	25.5	26.5	
Combined level, L_2	14.5	6.5	27.5	25.5	26.5	31.5
Criterion						35

* Where the levels are given to greater accuracy than the nearest dB, either round the numbers to the nearest whole dB and use the method given on page 7, or use the following formula to calculate the composite sound level:

$$L_{comp} = 10 \times \log (\text{antilog } L_1/10 + \text{antilog } L_2/10 + \text{etc})$$

Table A3

	Octave band centre frequency (Hz)					
	125	250	500	1000	2000	'A'
Calculate the level in the garden of plot 80						
Measured noise level, worst half-hour at 12 m	66	51	66	61	61	
A-weighting (See page 7)	−16	−9	−3	0	+1	
A-weighted level	50	42	63	61	62	67
Distance correction (3 dB per doubling)	−6	−6	−6	−6	−6	
Angle of view (See previous example)	−7	−7	−7	−7	−7	
Barrier attenuation (path difference: 1.5 m)	−13	−16	−19	−22	−26	
Received level, L_2	24	13	31	26	23	34
Criterion						65

Factory noise calculations

Brief 7

Use the results of a noise measurement survey on site to estimate the factory noise levels on plot numbers 30 and 52 in terms of the criteria set by the local authority (see pages 78 and 79). How can the future residents be protected from factory noise?

Data required

Plant located on the factory roof is the main source of noise in this case. The sound level on site does not fall off uniformly with distance, because the area of site closest to the factory is shielded from the noisy plant by the edge of the roof.

The only reliable method to estimate sound levels on site is by measurement. The plant runs 24 hours a day, so the best time to measure is during the night when the noise from other sources around the site has fallen to a minimum. The noise level does not vary with time, so short-duration measurements can establish the L_{Aeq} value over any reference time period (see page 7).

Mark out a grid on site and measure the sound level at each point. Interpolate between the grid points to draw noise contours on the site plan. Figure A14 shows the individual sound level measurements and the resulting noise contours in this case. The maximum sound level measured is 50 dB L_{Aeq}, close to the centre of the site.

Plot 30

Plot 30 has an unobstructed view of the factory (see Figure A15). Therefore, the noise level will be similar to that predicted by the measurement survey, 48 dB L_{Aeq}. This is below the criterion level of 65 dB $L_{Aeq,0.5h}$ for gardens.

Plot 52

Plot 30 provides some shielding for plot 52 (see Figure A15). Use Figures A16 and A17 (Figures 25 and 26 on page 22) to estimate the attenuation provided. The roof ridge on plot 30 coincides approximately with the line of sight between source and plot 52. (The roof is at 'grazing incidence' to the soundpath). This represents a path difference of 0 m which gives 5 dB attenuation at all frequencies (see Figure A17). Therefore, plot 52 will receive a noise level 5 dB lower than the level of 49 dB established by the measurement survey, 44 dB L_{Aeq}. This is below the criterion level of 65 dB $L_{Aeq,0.5h}$ for gardens.

Noise control measures

Barriers on the ground are of little use on sites where the noise source is so high above ground level. The best method for reduction would be to treat the source, but this is often outside the control of the site owner and the designer. This leaves only the possibility of providing the necessary sound insulation in the building envelope. The next example considers the design measures necessary to restrict factory noise to the criterion level of 35 dB $L_{Aeq,8h}$ internally at night.

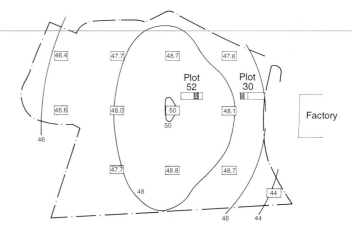

Figure A14

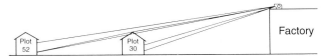

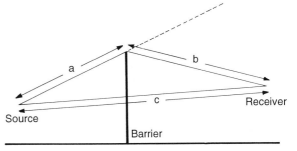

Figure A15

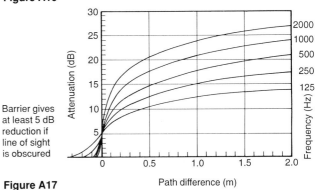

Path difference = a + b – c

Figure A16

Figure A17

Barrier gives at least 5 dB reduction if line of sight is obscured

Brief 8

Design the building envelope of the house on plot 30 to restrict factory noise inside the dwelling to no more than 35 dB $L_{\text{Aeq,8h}}$ at night.

Data required

The single-figure methods of calculation given on page 30 will not be accurate enough for these calculations because they rely on the source being typical road traffic. In this case, the data are required in octave bands. A typical spectrum has been measured during the site noise survey.

Formula

Noise from a single item of plant behaves like a point source, as long as the observer is not too close (see page 8). The following formula (which is derived in the box opposite) applies:

$$L_2 = L_1 - R + 10 \log (S/A) + 6 \text{ dB}$$

where

$L_1 = $ Average sound-pressure level close to the facade but not including the reflection effect of the facade (dB). (Use the result of the noise survey given in Table A4.)

$L_2 = $ Received sound level in the room (dB)

$R = $ Sound reduction index (dB). (See Figures 45 and 46, pages 30 and 31)

$S = $ Surface area, room facade element (m^2). Assume initially that the window = 1.5 m^2, opening area = 0.5 m^2, wall area = 5.5 m^2

$A = $ Absorption in the room (m^2). This will depend on room furnishings. Assume 10 m^2 at all frequencies, which is a reasonable assumption for a furnished room.

Calculation

Details of the calculation are shown opposite. If the window contains 4 mm glazing and is tightly closed, the internal noise level will be 19 dB(A). This easily meets the local authority criterion. If the window opening is 0.5 m^2, the internal noise level will be 41 dB(A), 6 dB(A) above the local authority criterion. The transmission through the wall and the window are negligible compared with the opening. The internal noise level depends on the area of the opening; every time the area is halved, the internal level will decrease by 3 dB. Therefore, the area of the opening should be reduced to one-quarter of its original area (to approximately 0.125 m^2) to meet the local authority criterion of 35 dB(A).

Result

Noise from the nearby factory will not exceed 35 dB $L_{\text{Aeq,8h}}$ inside the dwelling on plot 30 if the window-opening area does not exceed approximately 0.125 m^2.

Derivation of formula

On page 13, the following formula is given to determine the sound reduction index of a facade using a loudspeaker:

$$R_\theta = L_1 - L_2 + 10 \log \left(\frac{4 \times S \times \cos \theta}{A} \right) \text{ dB}$$

where $\theta = $ the angle of incidence in degrees and all other terms are defined in the text.

Rearranging the formula to obtain L_2 gives:

$$L_2 = L_1 - R_\theta + 10 \log \left(\frac{4 \times S \times \cos \theta}{A} \right) \text{ dB}$$

The highest value of L_2 will occur when $\theta = 0$ and $\cos \theta = 1$. Therefore, the worst-case formula would be:

$$L_2 = L_1 - R + 10 \log \left(\frac{4 \times S}{A} \right) \text{ dB}$$

But $10 \log 4 = 6$ dB, therefore:

$$L_2 = L_1 - R + 10 \log (S/A) + 6 \text{ dB}$$

Table A4

	Octave band centre frequency (Hz)					
	125	250	500	1000	2000	'A'
Measured noise level (Away from reflecting objects)	51	47	44	42	42	
A-weighting (See page 7)	−16	−9	−3	0	+1	
A-weighted level	35	38	41	42	43	48
Calculate the reduction in level through each element						
R wall, brickwork	−41	−45	−50	−54	−59	
$10 \times \log (S/A)$ ($S = 5.5$ m^2, $A = 10$ m^2)	−2.6	−2.6	−2.6	−2.6	−2.6	
+ 6	+6	+6	+6	+6	+6	
Sound level received through wall only	−2.6	−3.6	−5.6	−8.6	−12.6	
R, window	−20	−23	−26	−29	−29	
$10 \times \log (S/A)$ ($S = 1.5$ m^2, $A = 10$ m^2)	−8.2	−8.2	−8.2	−8.2	−8.2	
+6	+6	+6	+6	+6	+6	
Sound level received through window only	12.8	12.8	12.8	10.8	11.8	
R, open window (assumed)	0	0	0	0	0	
$10 \times \log (S/A)$ ($S = 0.5$ m^2, $A = 10$ m^2)	−13.0	−13.0	−13.0	−13.0	−13.0	
+6	+6	+6	+6	+6	+6	
Sound level received through opening only	28	31	34	35	36	
Combine the levels*						
Via wall	−2.6	−3.6	−5.6	−8.6	−12.6	
Via window	12.8	12.8	12.8	10.8	11.8	
Via opening	28	31	34	35	36	
Combined level, L_2 through wall and window	12.8	12.8	12.8	10.8	11.8	19.3
Combined level, L_2 wall, window and opening	28	31	34	35	36	41
Criterion						35

* Where the levels are given to greater accuracy than the nearest dB, either round the numbers to the nearest whole dB and use the method given on page 7, or use the following formula to calculate the composite sound level:

$$L_{\text{comp}} = 10 \times \log (\text{antilog } L_1/10 + \text{antilog } L_2/10 + \text{etc})$$

Dairy noise calculations
Brief 9

Use the results of noise measurements made on site close to the dairy to compute the external $L_{Aeq,0.5h}$ levels at plot 28 for comparison with local authority criteria (see pages 78 and 79).

Data required

Noise measurements have been made between 5 am and 8 am. During this period, the tractor units of delivery lorries start up, reverse to hitch up their loaded trailers, and depart. Over the same period, milk floats are continuously loaded at the loading bay (see Figure A18). Noise measurements are made at convenient distances for three activities:

● Lorry tractor unit start-up

Distance:	10 m from microphone, 7 m from plot 28
Duration:	up to 20 seconds for winter starting
Number:	12 lorry start-ups per hour
Level:	85 dB $L_{Aeq,T}$ where $T = 20$ seconds

● Lorry departure

Distance:	20 m from microphone, 13 m from plot 28
Duration:	2 minutes
Number:	12 lorries depart per hour
Level:	70 dB $L_{Aeq,T}$ where $T = 2$ minutes

● Loading floats

Distance:	20 m from microphone, 19 m from plot 28
Duration:	30 minutes
Level:	65 dB $L_{Aeq,T}$ where $T = 30$ minutes

Calculation

Lorry start-up and departure
The calculation procedure involves four steps:

1 Convert from L_{Aeq} to L_{AE} (see box for formula)
   Start-up: 85 dB $L_{Aeq,20s}$ $85 + 10 \log (20)$ dB
                           98 dB L_{AE} (at 10 m)

   Similarly, departure noise   91 dB L_{AE} (at 20 m)

2 Correct the L_{AE} values for distance (Figure A19*)
   Start-up measured at 10 m,   Reduction = 20 dB
   Distance to plot 28 = 7 m,   Reduction = 17 dB
                                Difference = 3 dB

   Start-up level at plot 28 = $98 + 3 = 101$ dB L_{AE}

   Similarly, departure level, plot 28 = 95 dB L_{AE}

3 Combine the L_{AE} values
   There are six start-ups at 101 dB L_{AE} and six departures at 95 dB L_{AE} in each half-hour. Combine these, using the rules for addition of dB given on page 7 or use the following method:

   Six times 101 dB = $101 + 10 \log 6 = 101 + 8 = 109$ dB
   Six times 95 dB = $95 + 10 \log 6 = 95 + 8 = 103$ dB
   109 and 103 (difference 6, add 1) = 110 dB

4 Convert from L_{AE} to $L_{Aeq,0.5h}$ (see box)

   110 dB L_{AE} $= 110 - 10 \log (30 \times 60)$
                     $= 110 - 33$
                     $= 77$ dB $L_{Aeq,0.5h}$

Loading bay
Noise from the loading bay has been measured as a half-hour L_{Aeq}. The distance correction (from 19 m to 20 m) has a negligible effect on the level. Therefore, the level from the loading bay is 65 dB $L_{Aeq,0.5h}$. This has an insignificant effect on the overall L_{Aeq}.

Result of the calculation

At plot 28 the external noise level which can be attributed to the nearby dairy is 77 dB $L_{Aeq,0.5h}$. This exceeds the local authority criterion of 65 dB $L_{Aeq,0.5h}$.

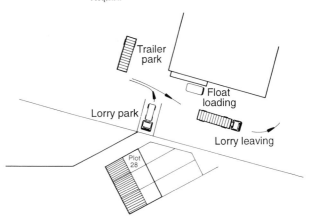

Figure A18

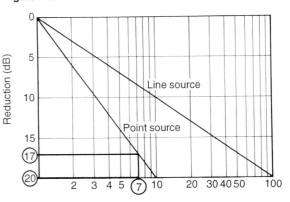

Figure A19

Converting between $L_{Aeq,T}$ and L_{AE}

The formula for obtaining L_{Aeq} values from L_{AE} values given on page 11 simplifies to the following:

$$L_{Aeq,T} = 10 \log \frac{\text{antilog } L_{AE}/10}{t} \text{ dB}$$

where t is the total time in seconds.
Therefore,

$$L_{Aeq,T} = 10 \log (\text{antilog } L_{AE}/10) - 10 \log (t)$$

Therefore,

$$L_{Aeq,T} = L_{AE} - 10 \log (t)$$

and:

$$L_{AE} = L_{Aeq,T} + 10 \log (t)$$

---

* Figure A19 is based on the following formula:

   Distance attenuation = $20 \log (d/r)$ dB

where

   d = source–receiver distance (m)

   r = reference distance (m)
   (Source–microphone distance in this example)

Brief 10

Calculate the noise level in the garden of plot 28 if a 2 m barrier is erected on the boundary of the site, close to the dairy. How can the requirements of the local authority be achieved?

Data required

Figure A20 shows a section through plot 28 and the dairy. Use the method given on page 22 to assess the barrier attenuation for each source. Combine the levels to obtain the overall level in the garden of plot 28.

Calculations

Barrier attenuation for lorry start-up
(Figure A21)

$$a = \sqrt{1.2^2 + 0.8^2} = 1.44 \text{ m}$$

$$b = \sqrt{5.4^2 + 0.5^2} = 5.42 \text{ m}$$

$$c = \sqrt{6.6^2 + 0.3^2} = 6.61 \text{ m}$$

Path difference $(a + b - c) = 0.25$ m

Use Figure A22 to obtain the barrier attenuation. No frequency information is given. Use the 125 Hz curve, which gives the most pessimistic result.

Result: barrier attenuation = 9 dB.

Check the barrier attenuation for departure and loading, using the dimensions given on Figure A20. Values of 7 dB and 6 dB should be obtained.

Garden noise calculation

	Start	Depart	Loading
L_{AE}, plot 28, no barrier	101	95	
Six events (+ 10 log 6)	+8	+8	
10 log (30 × 60) (see box)	–33	–33	
$L_{Aeq,0.5h}$ for each activity	76	70	65
Barrier effect	–9	–7	–6
$L_{Aeq,0.5h}$, garden, plot 28:	67	63	59

Combined level, garden   69 dB $L_{Aeq,0.5h}$
Criterion, garden   65 dB $L_{Aeq,0.5h}$

Result of the calculation

Noise from the dairy exceeds the criterion for gardens set by the local authority. Consider altering the site layout to remove private gardens from the site boundary near to the dairy. Figure A23 shows a possible layout. The high-rise blocks which have no private grounds are relocated towards the northern boundary. The surrounding area is used for car parking.

Alternatively, consider placing single-aspect terraced housing close to the northern site boundary, with private gardens on the south side where they would benefit from the barrier effect of the terrace.

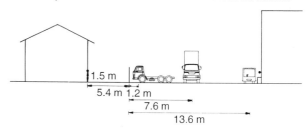

Figure A20

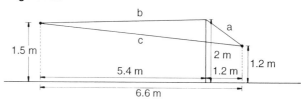

Figure A21

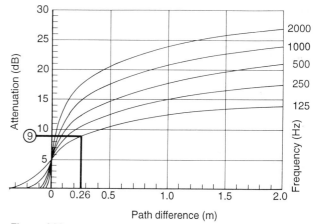

Figure A22

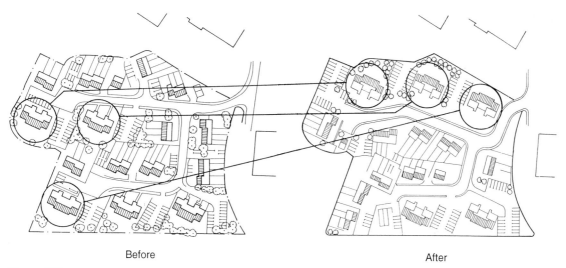

Before   After

Figure A23

119

Brief 11

Calculate the noise level 1 m from the second-floor windows of the high-rise block closest to the dairy. How much sound insulation is required from the building envelope, and how can it be achieved in practice?

Data required

Figure A24 shows a section through the dairy and the high-rise block. Use the method given on page 22, to assess the barrier attenuation for each source. Combine the levels to obtain the overall level at the building facade.

Calculations

There is a clear line of sight between the lorry departure position and the second-floor windows of the block. The barrier still provides some attenuation, however. The barrier attenuation can be assessed for each source, using the dimensions given on Figure A24 together with Figure A22. The following values should be obtained:

Start-up 7 dB, Departure 3 dB, Loading 1 dB

Details of the facade level calculation are given in Table A5. Resulting level at facade = 70 dB $L_{Aeq,0.5h}$

Building envelope sound insulation

Criterion inside dwelling = 35 dB $L_{Aeq,8h}$

The dairy noise must be converted to an 8-hour L_{Aeq} before it can be compared directly with the criterion.

The night period is 11 pm to 7 am (8 hours). The dairy operates between 5 am and 8 am, which includes 2 hours during the night period. Therefore, the night-time dairy noise is 70 dB $L_{Aeq,2h}$

Average the sound energy over the reference time period to obtain the 8-hour L_{Aeq}. Every time the reference time period is doubled, the equivalent continuous sound energy is halved. In dB terms, this is represented by a decrease of 3 dB (10 log 2):

$$2\text{-hour } L_{Aeq} = 70 \text{ dB}$$
$$4\text{-hour } L_{Aeq} = 67 \text{ dB}$$
$$8\text{-hour } L_{Aeq} = 64 \text{ dB}$$

Therefore, dairy noise = 64 dB $L_{Aeq,8h}$

(If other noise sources are present during this period, add them to this before assessing the sound-insulation requirements).

Criterion inside dwelling = 35 dB $L_{Aeq,8h}$
Sound insulation needed = 29 dB(A)
(See Table A6.)

Construction

There is no frequency information available. Use the single-figure method given on page 30 to decide what the building envelope construction should be. Assume that the window in question occupies one-quarter of the area of a masonry wall (see Figure A25 and lower box).

This approximate calculation suggests that the sound insulation requirement can be met by closed 4 mm single glazing. It will, however, be necessary to provide ventilation by some means other than the window.

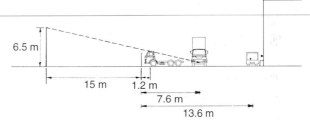

Figure A24

Table A5

Facade level calculation

	Start	Depart	Loading
L_{AE} single event	98	91	
Six events (+ 10 log 6)	+8	+8	
10 log (30 × 60)	−33	−33	
$L_{Aeq,0.5h}$	73	66	65
Measurement distance, r (m)	10	20	20
Distance to facade, d (m)	16.2	22.6	28.6
20 log (d/r)	−4	−1	−3
Barrier attenuation	−7	−3	−1
Facade reflection	+3	+3	+3
Level at facade	65	65	64

Combined level at facade 70 dB $L_{Aeq,0.5h}$

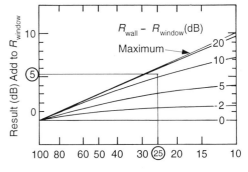

Window area as a percentage of total wall area

See Table 14 for typical R values

Figure A25

Table A6

Estimate of sound insulation of building facade

	$R_{A(traffic)}$
Brick wall containing Noise Insulation Regulations ventilator	39
4 mm single-glazing, good fit	24
R (wall) − R (window)	15
25% window area, add to R (window) (See Figure A25)	5
Composite sound insulation	29
Required sound insulation	29

Appendix B
Designers' checklists

Solid masonry walls

Set the MASS high enough.

As the mass of the wall is reduced, the CRITICAL FREQUENCY increases to a point where its effects are more serious (see page 14).

Specify that bricks be laid frog-up, and that all joints be fully filled with mortar.

Specify a plaster finish to seal acoustically-porous blocks.

Dry linings and other panel finishes can cause reduced low-frequency performance as a result of the MASS-AIR RESONANCE, particularly on dense concrete walls (see page 15).

Observe the guidance given on pages 37 and 38 regarding window spacing in external flanking walls.

In the separating wall, avoid areas of reduced MASS, such as electrical sockets, which should not be installed back-to-back.

Joists should preferably span parallel to the separating wall.

Masonry walls between isolated panels

Specify the elements to achieve adequate MASS.

Specify an appropriate cavity width.

Do not permit anything to bridge the cavities.

Specify that all joints be fully filled with mortar, and that bricks be laid frog-up.

Specify that all joints at the perimeter of the lightweight panel be well sealed, for example with tape, caulking or coving.

Joists should span parallel to the separating wall. If this is not possible, use joist hangers.

Suspended concrete floor with a soft covering

Set the MASS high enough by specifying the concrete mix and thickness, or by appropriate selection of precast elements.

In thin slabs, the CRITICAL FREQUENCY will rise to a point where its effects are more serious (see page 14).

Specify that all joints be well formed and free from honeycombing.

Keep services penetrations to a minimum, and encase pipes.

Ceilings which enclose a small airgap can cause reduced low-frequency performance as a result of the MASS-AIR RESONANCE (see page 15).

All junctions should be well filled and sealed; check the specification of all junction details (see page 51).

Build the floor in on all sides.

Cavity masonry walls

Set the MASS high enough.

Specify an appropriate cavity width.

Specify only butterfly wire ties, and do not space them too closely together.

Timber frames with absorbent material

Do not specify anything which bridges the cavity other than necessary metal straps at the recommended spacing, and suitably-designed cavity closers.

Avoid services penetrations through the linings.

If socket outlets must be located on a separating wall, specify that the lining enclose the boxes, and do not position them back-to-back.

Specify that all perimeter joints be well sealed, for example with mastic, jute scrim or coving.

Suspended concrete floor with a floating layer

Set the MASS high enough by specifying the concrete mix and thickness, or by appropriate selection of precast elements.

In thin slabs, the CRITICAL FREQUENCY will rise to a point where its effects are more serious (see page 14).

Specify that all joints be well formed and free from honeycombing.

Keep services penetrations to a minimum, and encase pipes.

Specify no bridging between floating layer and concrete base, and do not run services under the floating floor.

A floating timber raft can cause a dip in performance as a result of the MASS-AIR RESONANCE (see page 15). Ensure that the battens are deep enough.

Ceilings which enclose a small airgap can cause reduced low-frequency performance as a result of the MASS-AIR RESONANCE (see page 15).

Check the specification of all junction details.

Check with the resilient layer manufacturer that the product is suitable for long-term wear under a timber raft construction.

Specify that the resilient layer be dry when installed. Dampness affects its resilience.

Build the floor into the walls on all sides.

Suspended timber-joist floors

Specify the density and thickness of the mineral-wool resilient layer.

Check with the manufacturer that the product has long-term resilience under the concentrated loadings imposed by a domestic ribbed timber floor.

Lightweight masonry flanking walls can cause flanking transmission problems unless an independent wall lining is adopted.

Specify that pugging be dry and laid on plastic sheeting.

Keep services to a minimum, and encase pipes.

Specify that there be no bridging between floating layer and base.

Check the specification of all junction details.

Suspended timber-joist floors in conversions

Treat existing thin masonry walls to control flanking transmission. (See page 61 for measures to control flanking transmission.)

Specify that pugging be dry and laid on plastic sheeting.

Specify that there be no contact between an independent ceiling and the existing ceiling.

Specify that there be no bridging of the resilient layer in any floating floor construction.

Check with the manufacturer that the product has long-term resilience under the concentrated loadings imposed by a domestic ribbed timber floor.

Specify sealing between the perimeter of the ceiling and the surrounding constructions.

Walls in conversions

Inspect the existing construction for gaps, particularly between the joists of intermediate floors, and specify that they be sealed.

Specify that there be no physical contact between an independent lining and the existing construction.

Specify good sealing around the perimeter of an independent lining.

Appendix C
Site inspection checklists

Check with job specification before inspecting.

Solid masonry walls

* Ensure that the bricks or blocks on site are of the correct density.

    Ensure that the joints are fully filled with mortar, and that bricks are laid frog-up.

    Check all junctions with surrounding constructions to ensure that there is an airtight seal.

* Look for areas of reduced mass and ensure that these are airtight and that the reduction in mass is minimal.

    Check the means of support for joists on the separating wall. Joist hangers should be used.

* Inspect the separating wall in the roof space and where it passes through an intermediate floor.

* Ensure that the plaster finish thickness is as specified.

Masonry walls between isolated panels

* Ensure that the bricks or blocks on site are of the correct density.

    Ensure that all joints are fully filled with mortar, and that bricks are laid frog-up.

    Check all junctions with surrounding constructions to ensure that there is an airtight seal.

* *Most important:* ensure that the lightweight panels do not make physical contact with the masonry core, and that the specified cavity width is maintained.

* Inspect the separating wall, in the roofspace and where it passes through an intermediate floor.

Suspended concrete floor with a soft covering

* Check the thickness of the suspended concrete slab and the mix for in-situ concrete, in particular the quantity and type of aggregate used.

* Check all junctions with surrounding constructions, and ensure that all joints are well formed and free from honeycombing.

* Ensure that services penetrations are no larger than necessary, and that any pipes are properly wrapped.

* Ensure that the floor is built-in on all sides, if specified.

Cavity masonry walls

* Ensure that the bricks or blocks on site are of the correct density.

* Ensure that only butterfly ties are being used, and that they are correctly spaced.

* Ensure that the specified cavity width is maintained, and that the cavities are not bridged by any unspecified material.

    Ensure that the joints are fully filled with mortar, and that bricks are laid frog-up.

    Check all junctions with surrounding constructions to ensure that there is an airtight seal.

* Check that areas of reduced mass are airtight, and that the reduction in mass is minimal.

* Inspect the separating wall, in the roof space and where it passes through an intermediate floor.

* Ensure that the plaster finish thickness is at least as specified.

Timber-frame walls with absorbent material

* *Most important:* ensure that the specified cavity width is provided, and that no unspecified items bridge the cavity.

    Ensure that there are no unspecified penetrations of the linings.

    Ensure that any socket outlets in the separating wall are enclosed by the cladding material and not located back-to-back.

* Ensure that the absorbent quilt is of the specified thickness, and has been fixed in the specified position.

* Ensure that the plasterboard linings are of the specified thickness, the joints staggered and the perimeter well-sealed, in accordance with the specification.

Suspended concrete floor with a floating layer

* Check the thickness of the concrete base and the mix for in-situ concrete, in particular the quantity and type of aggregate used.

* Check all junctions with surrounding constructions, and ensure that all joints are well formed and free from honeycombing. Joints in beam/block and plank floors should be filled.

    Ensure that the resilient quilt is dry on installation.

* *Most important:* undertake frequent and detailed inspections to ensure that there are no elements bridging between the floating layer and the base floor.

    Ensure that the base floor is level and smooth before the resilient layer is laid.

* *Most important:* check the grade and thickness of the resilient layer on site.

* Ensure that the resilient layer is turned up at the edges to ensure that the edge of the floating layer does not touch the flanking construction.

    Ensure that there is a narrow gap left between the bottom of the skirting board and the floating layer. Fill this gap only with a permanently-soft material.

* Ensure that the floor is built-in on all sides, if specified.

Suspended timber-joist floors

* *Most important:* ensure that the resilient layer on site is as specified.

* Ensure that any pugging is dry and installed to the correct thickness on a plastic sheet.

  Ensure that all layers in the construction are laid without gaps.

* Ensure that the resilient layer is turned up at the edges or that a mineral-fibre or plastics foam strip fills the gap between the edge of the floating floor and the walls.

  Ensure that there is a narrow gap between floating floor and skirting. Fill this gap only with a permanently-soft material.

* *Most important:* undertake frequent and detailed inspections to ensure that there are no nails or any other elements, such as services pipes and conduits, bridging between the floating floor and the base.

Suspended timber-joist floors in conversions

* *Most important:* ensure that there is no physical contact between an independent ceiling and the existing construction.

* *Most important:* ensure that any resilient layer on site is as specified.

* *Most important:* ensure that there is no bridging of any resilient layer in the construction, for example by nailing through a floating floor.

* Ensure that any resilient layer is turned up at the edges and that there is a gap between any floating layer and the skirting.

  Ensure that all layers in the construction are laid without gaps.

Walls in conversions

Most important: ensure there is no physical contact between an independent lining and the existing wall.

Most important: where resilient bars are used, ensure that the only contact between the new plasterboard lining and the studs is via the resilient bars.

Ensure that all layers in the construction are constructed without gaps, and that there are no gaps in the existing layer.

Ensure that any gaps between the joists of an intermediate floor are filled with plasterboard or timber blocks to a good tight fit.

Appendix D
Method for calculating mass

Notes

Where a mass is specified for walls or floors, it is expressed in kg/m².

The mass may be obtained from actual figures given by the manufacturers, or it may be calculated by the method given here. To calculate the mass of a masonry leaf, use the appropriate formulae from Table D1. These formulae are not exact, but are accurate enough for this purpose. For co-ordinating heights other than those given in the table, use the formula for the nearest height.

Densities of bricks or blocks (at 3% moisture content) may be taken from a current certificate of the British Board of Agrément (BBA) or from a European Technical Approval Certificate (ETA, as defined in the Construction Products Directive of the European Community) or from the manufacturer's literature, in which case the building control authority may ask for confirmation, for example that the measurement was done by an accredited test house, if the wall separates dwellings.

The quoted density of bricks or blocks is normally the apparent density, in other words the weight divided by the volume, including perforations, voids or frogs. This is the density appropriate to the formulae for the nearest height. Include any finish of plaster, render or dry lining in calculating the mass unless otherwise stated.

A mortar joint of 10 mm and a dry set mortar density of 1800 kg/m³ are the assumed values. Values within 10% of these figures are acceptable.

For in-situ concrete or screeds calculate the mass by multiplying the density (kg/m³) by the thickness in metres. For slabs or composite floor bases divide the total mass of the element (kg) by the plan area of the element (m²).

Table D1
Formulae for calculation of wall leaf mass

Co-ordinating height of masonry course	Formula to be used
75	$M = T(0.79D + 380) + NP$
100	$M = T(0.86D + 255) + NP$
150	$M = T(0.92D + 145) + NP$
200	$M = T(0.93D + 125) + NP$

Where

$M =$	Mass of 1 m² of leaf (kg/m²)
$T =$	Thickness of masonry (m) (Unplastered thickness)
$D =$	Density of masonry units in kg/m³
$N =$	Number of finished faces (if no finish $N = 0$, if finish on one side only $N = 1$, if finish on both sides $N = 2$)
$P =$	Mass of 1 m² of wall finish in kg/m². (see below)

Finishes

Mass of plaster (assumed thickness 13 mm)

Cement render	29 kg/m²
Gypsum	17 kg/m²
Lightweight	10 kg/m²
Plasterboard	10 kg/m²

Appendix E
Glossary of acoustic terms

A-weighting
Frequency response curve which approximates to the sensitivity of the ear (page 7)

Absorption
Conversion of sound energy to heat (page 9)

Absorption coefficient
The proportion of sound energy absorbed, expressed as a number between 0 and 1

Airborne sound
Sound energy which is transmitted from a source by excitation of the surrounding air (page 12)

Ambient noise
Totally encompassing sound in a given situation at a given time

Background noise
Usually described in terms of the L_{A90} level: the level exceeded for 90% of the time

Critical frequency
The frequency at which the wavelength of sound in the air coincides with the wavelength of the associated vibration in a wall, floor or partition (page 14)

Damping
Conversion of vibrational energy into heat, which makes a body a less efficient sound radiator

Decibel (dB)
Unit of sound-pressure level (page 6)

Diffraction
The deflection of a sound wave caused by an obstruction in a medium (page 9)

Diffuse field
A space in which the sound-pressure level is constant for all positions

Equivalent continuous A-weighted sound-pressure level (L_{Aeq})
The level of a notional steady sound which, over a defined period of time, T, would deliver the same A-weighted sound energy as the fluctuating sound (page 7)

Filter set
A device used in sound-level meters to allow frequency analysis, for example in octave bands of a sound (page 7)

Flanking transmission
Sound transmitted between rooms via elements which are common to both rooms, other than the element which separates them (page 15)

Free field
A space without sound-reflecting surfaces

Frequency
The number of times in one second that a cyclic fluctuation repeats itself (page 6)

Hertz (Hz)
Unit of frequency, cycles per second (page 6)

Impact sound
Sound energy resulting from direct impacts on the building construction

Impact sound-pressure level (L_i)
The sound-pressure level in a room, resulting from impacts on the floor above generated using a standardised impact sound source (page 12)

Isolation
The introduction of a discontinuity between two elements in an energy transmission chain (page 14)

L_{AN}
The A-weighted sound-pressure level which is exceeded for N% of the time (page 10)

Level difference (D)
The arithmetic difference between the sound-pressure level in the source room and that in the receiving room (page 12)

Line source
A sound source which can be idealised as a line in space (page 8)

Mass law
The empirical relationship between partition mass and sound insulation (page 14)

Mean sound-insulation values (for example R_m)
The arithmetic average of the 16 sound-insulation values between 100 Hz and 3150 Hz (page 30)

Noise criteria (NC)
A set of octave band sound-pressure level curves used for specifying limiting values for building services noise

Noise rating (NR)
Similar to NC.

Noise reduction coefficient (NRC)
The mean absorption coefficient of a material, averaged over four octave bands: 250, 500, 1000 and 2000 Hz.

Noise and number index (NNI)
An index formerly used to describe aircraft noise around large airports (page 11)

Octave
A frequency ratio of 2 (page 6)

Pascal
A unit of pressure corresponding to a force of 1 newton acting on an area of 1 m^2 (page 6)

Pink noise source
A source which delivers all frequencies with constant energy per octave or one-third octave band

Point source
A sound source which can be idealised as a point in space (page 8)

$R_{A(traffic)}$
Value obtained using a procedure given in this manual (page 30) to estimate the difference between external and internal traffic noise levels in dB(A)

Rating level
The $L_{Aeq,T}$ value associated with the industrial noise source under investigation. Used in British Standard BS 4142 (page 11)

Reference time interval
The specified interval over which an equivalent continuous A-weighted sound-pressure level is determined (page 7)

Reflection
The phenomenon by which a wave is returned at a boundary between two media (page 9)

Resonant frequency
A frequency at which the response of a vibrating system to an input force reaches a maximum (page 15)

Reverberation time (T)
The rate of decay of sound in a room (page 12)

Root mean square (RMS)
The square root of the arithmetic average of a set of squared instantaneous values

Sabin
A measure of sound absorption. One sabin equals 1 m^2 of perfectly-absorbing surface

Sound power
The total sound energy radiated by a source per second

Sound transmission class (STC)
A single-figure rating of sound reduction index. Similar to R_w. (USA)

Sound exposure level (L_{AE}, L_{AX} or SEL) The level which, if maintained constant for a period of one second, would deliver the same A-weighted sound energy as a given noise event (page 7)

Sound pressure
A dynamic variation in atmospheric pressure. The pressure at a point in space minus the static pressure at that point (page 6)

Sound-pressure level
The fundamental measure of sound pressure defined as:
$SPL = 10 \log (p/p_o)^2$ dB, where p is the RMS value of sound pressure in pascals, and p_o is 0.00002 pascals (page 6)

Sound reduction index (R)
Ratio of the sound energy emitted by a material to the energy incident on the opposite side (page 12)

Standardised values
Sound-insulation values which have been adjusted to a receiving room reverberation time of 0.5 seconds (pages 12 and 13)

Structureborne sound
Sound energy which is transmitted directly into the building construction from a source which is in contact with it (page 32)

Wavelength
In a cyclic fluctuation, the distance between the ends of a single complete cycle (page 6)

Weighted sound-insulation values
(for example R_w)
Sound insulation values obtained over the one-third octave band frequency range 100 Hz to 3150 Hz are turned into a single-figure weighted value using the procedure given in British Standard BS 5821 (pages 16 and 17)

White noise source
A source which delivers all frequencies with constant energy per unit of frequency

Appendix F
Information sources

AACPA
Autoclaved Aerated Concrete Products Association,
7 Buckingham Gate,
London SW1E 6JS.
Telephone   071 630 7574

ANC
Association of Noise Consultants,
6 Trap Road,
Guilden Morden
Hertfordshire SG8 0JE.
Telephone   0763 852958

BCA
British Cement Association,
Wexham Springs,
Slough,
Buckinghamshire SL3 6PL.
Telephone   0753 662727

BDA
Brick Development Association,
Woodside House,
Winkfield,
Windsor,
Berkshire SL4 2DX.
Telephone   0344 885651

BG
British Gypsum Ltd,
East Leake,
Loughborough,
Leicestershire LE12 6QJ.
Telephone   0602 456123

BRE
Building Research Establishment,
Garston,
Watford,
Hertfordshire WD2 7JR.
Telephone   0923 894040

CIRIA
Construction Industry Research and Information Association,
6 Storey's Gate,
Westminster,
London, SW1P 3AU.
Telephone   071 222 8891

EURISOL-UK
Mineral Wool Association,
39 High Street,
Redbourn,
Hertfordshire AL3
Telephone   0582 794624

PG
Pilkington Glass Ltd.
Prescot Road,
St Helens,
Merseyside, WA10 3TT
Telephone   0744 28882

TRADA
Timber Research and Development Association,
Stocking Lane,
Hughenden Valley,
High Wycombe,
Buckinghamshire HP14 4ND .
Telephone   0240 243091

Appendix G
Bibliography

1   **Department of the Environment and the Welsh Office.** *The Building Regulations. Approved Document E: Resistance to the passage of sound* (1992 edition). London, HMSO, 1991.

2   *Building Regulations (Northern Ireland). Technical Booklets G and G1; Sound.* Belfast, HMSO, 1990.

3   *The Building Standards (Scotland) Regulations. Part H: Resistance to transmission of sound.* London, HMSO, 1990.

4   *Building and buildings. The Noise Insulation Regulations 1975.* Statutory Instrument 1975 No 1763.

5   **Department of the Environment and the Welsh Office.** *Calculation of road traffic noise.* London, HMSO, 1988.

6   **British Airports Authority.** *Heathrow noise insulation grants scheme* (amended June 1981). London, BAA, 1981.

7   **British Standards Institution.** Method of rating industrial noise affecting mixed residential and industrial areas. *British Standard* BS 4142:1990. London, BSI, 1990.

8   **British Standards Institution.** Method of measurement of sound insulation in buildings and of building elements. *British Standard* BS 2750:1980. Parts 1 to 8. London, BSI, 1980. (Under revision)

9   **British Standards Institution.** Rating the sound insulation in buildings and of building elements. *British Standard* BS 5821:1980. Parts 1 to 3. London, BSI, 1980.

10   **Department of the Environment.** Planning and noise. Circular 10/73. London, DOE, 1973.

11   **Department of the Environment.** Planning and noise (draft Planning Policy Guidance, issued for comment 1992). London, DOE, 1992.

12   **House of Commons.** Environmental Protection Act 1990. *Chapter* 43. London, HMSO, 1990.

13   **British Standards Institution.** Code of practice for sound insulation and noise reduction for buildings. *British Standard* BS 8233:1987. London, BSI, 1987.

14   **Noise Advisory Council.** A guide to measurement and prediction of the equivalent continuous sound level L_{eq}. London, HMSO, 1978.

Printed in the UK for HMSO Dd.8379491, 3/93, C20, 38938